T0178248

History of Physics

The Springer book series *History of Physics* publishes scholarly yet widely accessible books on all aspects of the history of physics. These cover the history and evolution of ideas and techniques, pioneers and their contributions, institutional history, as well as the interactions between physics research and society. Also included in the scope of the series are key historical works that are published or translated for the first time, or republished with annotation and analysis.

As a whole, the series helps to demonstrate the key role of physics in shaping the modern world, as well as revealing the often meandering path that led to our current understanding of physics and the cosmos. It upholds the notion expressed by Gerald Holton that "science should treasure its history, that historical scholarship should treasure science, and that the full understanding of each is deficient without the other." The series welcomes equally works by historians of science and contributions from practicing physicists.

These books are aimed primarily at researchers and students in the sciences, history of science, and science studies; but they also provide stimulating reading for philosophers, sociologists and a broader public eager to discover how physics research – and the laws of physics themselves – came to be what they are today.

All publications in the series are peer reviewed. Titles are published as both print- and eBooks. Proposals for publication should be submitted to Dr. Angela Lahee (angela.lahee@springer.com) or one of the series editors.

More information about this series at https://link.springer.com/bookseries/16664

Alessandro De Angelis

Galileo Galilei's "Two New Sciences"

for Modern Readers

 Springer

Alessandro De Angelis ⓘ
Dipartimento di Fisica e Astronomia
"Galileo Galilei"
Università di Padova
Padova, Italy

ISSN 2730-7549   ISSN 2730-7557   (electronic)
History of Physics
ISBN 978-3-030-71954-8   ISBN 978-3-030-71952-4   (eBook)
https://doi.org/10.1007/978-3-030-71952-4

To the memory of Antonio Favaro

All knowledge of reality starts from experience and ends in it. Propositions arrived at by purely logical means are completely empty as regards reality. Because Galileo saw this, and particularly because he drummed it into the scientific world, he is the father of modern physics—indeed, of modern science altogether.

Albert Einstein, "On the method of theoretical physics", Oxford, 1933

Preface by Ugo Amaldi

With great surprise I received in July 2019 from Alessandro De Angelis the draft of a modern version of Galileo's *Discourses and mathematical demonstrations related to two new sciences*; De Angelis asked me for an opinion on his work. A few sentences in his cover letter were saying: "Although the title contains the word 'mathematics', Galilei, like Newton, manipulated algebraic formulas only in a limited way and instead used geometry for his proofs. Notational mathematics, such as $F = ma$, and analytic geometry, were being developed at the same time as Galileo, and he did not use them. Furthermore, Galilei wrote in a somewhat 'baroque' way (excuse me for the expression) and his writings are difficult to understand. Understanding Galilei requires some knowledge of the classic Latin and Greek literature and a passion for physics, qualities not so common to find together. However, many could be enriched by knowing the art, intelligence, and beauty of his arguments, and by sharing the wonder that can often be encountered between the lines of his writings. For this reason, I decided to translate Galilei's *Discourses and mathematical demonstrations* in modern language and algebraic formulas to make him current and understandable by those who I imagine are 'modern' learned readers: curious, passionate about science, but unfortunately with little time to deepen the lexical, historical, and philosophical antiquities."

I was immediately reminded of an episode, unique in my life, happened thirty years earlier, in 1990, when for about ten years I had been the spokesperson of the international collaboration DELPHI, made up of about five hundred physicists from about twenty different countries. A year earlier, we had finished building a particle detector for CERN's electron-positron collider LEP and, having collected a lot of data, we were publishing our first scientific papers. The young Alessandro was a graduate student at the University of Padua, one of the thirty graduate students of the collaboration, with whom I had talked about physics, sometimes finding him highly educated for his age and open to new ideas. Entering my office with shyness, he placed on my desk a scientific note ready for publication, with all the necessary bibliographic indications, on a subject different from all those on which hundreds of much older and more experienced collaborators worked; when I read it I was struck by the clarity of the exposition and the completeness of the data analysis. That communication on the phenomenon of "intermittency," soon published by a prestigious magazine, is still one of the most referenced and interesting papers published by DELPHI.

A few years later I resigned from DELPHI spokesperson to deal with applications of hadron accelerators to cancer therapy and Alessandro left particle physics to work in astroparticle physics so that we did not have many opportunities to meet, even if I could follow in scientific journals the interesting results obtained with the MAGIC telescope at the international observatory of La Palma, a telescope of which he was one of the inventors. Later, in 2015, I found on my desk at CERN a 700-page volume, written with Mário Pimenta—who had been as well a graduate student in the DELPHI group in Lisbon many years earlier—and published by Springer under the title *Introduction to Particle and Astroparticle Physics.* Reading the last chapter, dedicated to *Astrobiology and the Relation of Fundamental Physics to Life,* was in particular a great intellectual pleasure. I was once again amazed by the quality and originality of his work.

When I went through the first version of the present book I experienced the same feelings of astonishment and intellectual pleasure. As he told me by phone shortly after his first email, De Angelis had been passionate about *Discourses and mathematical demonstrations* since high school, when on the sidelines he noted the translation into an algebraic language of the proofs based on geometry: "For me to demonstrate geometrically is a bit like looking at things from above, synthetically; demonstrating algebraically is like looking at them from below, analytically."

The last book published by Galilei, *Discourses and mathematical demonstrations...* is, in a sense, his first one, because from the very beginning of his teaching in Pisa Galilei started to collect, also with some help from his students, his notes on mechanics. Throughout his life, and in the Paduan period in particular, he continued filling notebooks on this subject, until he finished this book at a late age in 1638. The Preface to the *Discourses...* reveals the concern of the publisher Lodewijk Elzevir—that Galilei found in Holland with much effort—who feared it would not be taken into sufficient consideration since Galileo was famous for the publication, in 1632, of the *Dialogue concerning the two chief world systems,* and then writes (in the paraphrase/translation by De Angelis): "The divine and natural gifts of Galilei are clear in the present work where he shows to have discovered, through many labors and vigils, two entirely new sciences, and to have demonstrated them in a conclusive, i.e., mathematical, way. What is even more remarkable in this work is the fact that one of the two sciences deals with a very old subject, perhaps the most important in nature: [..] I refer here to motion. [..] The other science which he has also developed from its very foundations deals with the resistance which solid bodies offer to fracture by external forces, a subject of great utility, especially in the sciences and in the art of construction. [,,,] This book treats for the first time these two sciences and is full of conclusions to which, over time, others will be added by new thinkers. Moreover, through a large number of very clear demonstrations, the author paves the way for many new theorems that will be demonstrated by intelligent readers." History has taught that Lodewijk Elzevir's concerns about a possible lack of interest in Galilei's book had no basis.

The *Discourses and mathematical demonstrations related to two new sciences* is the seminal work of the scientific method, and reading this book is enlightening not only for physics students and professors but also for all science enthusiasts, and for anyone who wants to understand the history of human thought. The consideration, at the basis of this dialogue, that experiment and demonstration are the key tools for understanding nature, represents an imperishable message even in its apparent simplicity. The wonder at Galilei's persuasive demonstrations and the simple examples and experiments he proposed to support his arguments broaden the mind and nourish the culture of curious readers.

We must be truly grateful to Alessandro De Angelis who made this Galilei's book, which is at the foundation of all modern science, pleasant to read even for today's readers, accustomed to the use and abuse of the scientific culture of Wikipedia, and who gives with this work of his is a very important contribution to the understanding and interpretation of Galilei.

CERN, Geneva, Switzerland

Ugo Amaldi
Physicist, researcher and teacher
President Emeritus of the TERA
Foundation for Oncological Adrotherapy

Preface by Telmo Pievani

According to Galileo Galilei, the book of nature is written in mathematical language: more precisely, its "characters are triangles, circles, and other geometric figures." Alessandro De Angelis, four centuries later, translates Galilei's book of nature in algebraic terms. Whether you observe it synthetically from above or analytically from below, the revolutionary matter of the *Dialogues* you are about to read does not change. However, here is the bet, paraphrasing it makes it more readable, and its argumentative structure becomes clearer. Of course, laying hands on late Galileo's masterpiece and translating it into modern language is a difficult task, but here it is faced with the utmost seriousness.

There is a precedent. The decipherer of the secrets of stellar evolution, the 1983 Nobel Prize in Physics Subrahmanyan Chandrasekhar, in the last years of his life, between 1990 and 1995, had ventured into a similar work with Newton's *Principia*. He had rewritten the *Philosophiae Naturalis Principia Mathematica*, also in that case replacing geometric reasoning with formal mathematical notation, selecting the crucial moments, and extending the demonstrative passages. Newton's expert scholars, while applauding the attempt in itself, had however noted a series of interpretative distortions due to insufficient consideration of the historical context. The underlying problem lies indeed in the actualization, in the residual infidelity of each translation, and in the risk of introducing anachronisms. De Angelis did not let himself be dissuaded by such a precedent and brought to completion a project he had in mind since his juvenile studies. So here he rewrites for modern readers the *Discourses and mathematical demonstrations related to two new sciences concerning mechanics and local movements* by Galileo, which precedes the Newtonian *Principia* by fifty years and, by explicit admission of Newton, deeply inspires them.

However, there are some differences with the work of Chandrasekhar, all in favor of the *esprit de finesse* of De Angelis. Here the version is unabridged, except for a few and delicate reductions and additions: it is not a miscellany, therefore the arbitrariness of the selection is avoided and the work is returned to the reader in its entirety, including the additional day of dialogue on the force of percussion. There is notable attention to the history of criticism, to the context of the time and to the literality of the text, also in the use of the original drawings probably attributed to the hand of Galilei himself, at least for those of the first three days, as well as in the choice of adopting only the mathematical tools known at the time in Europe. So it is in some respects a version of the *Discourses* as Galileo himself could have written it, had he not made different choices on the basis of his knowledge. Furthermore, the language is paraphrased in an informal and cordial tone, with a refined and accurate set of notes related to style, content, history, and bibliography. Finally, the merit of De Angelis is to make all his methodological choices transparent in the Afterword. The result is a truly rigorous divulgation, which has also the interesting effect of making the *Discourses* more similar to the *Dialogue concerning the two chief world systems* published by Galileo six years earlier.

Indeed, this book also faces another challenge. We know that Galileo's prose, a model of Leopardi's, made Italo Calvino say in 1967 that he was the "greatest writer of Italian literature of all times," a combination of precision, evidence, and lyricism. We also know that this was not just a question of style. To counteract the obscurity and verbiage of the academic and ecclesiastical authorities, Galileo put in place a real strategy of cultural policy. He wrote in vernacular Italian to reach all who were curious enough to open up to the new vision of the cosmos, and perhaps capable of being excited by the unfolding of an open Universe and of a map of the world largely to be explored yet. The ideas of a new astronomy and new physics thus also became a theatrical tale and public debate. Yet, as De Angelis points out, when Galileo writes he is not always clear and linear.

Although being also written in dialogical and narrative form, the *Discorsi* in their original version are presented as a strange hybrid of vulgar and Latin, almost a step behind the *Dialogue*. They contain convoluted sentences, rather difficult paragraphs, passages that are not always explicit. Perhaps the rush of the last years, or the fears of Galilei after his trial, make the book difficult to read. Also, although the characters are the same as in the *Dialogue*, the roles the three play on stage are less intuitive. There are no more the peripatetic, the Copernican, and the connecting figure between the two, but different phases of Galilei's own scientific thought are dramatized, from youth to maturity. With a genial choice, the *Discorsi* thus become an entirely interior theater, the story of an intellectual parabola, a succession of hypotheses, discoveries, experiments, and demonstrations that are transferred from the scientist's head to the voices of the various characters. A scientific revolution is seen as it unfolds, from the inside.

Indeed, already in the *Dialogue*, if reread today, Simplicio can appear, rather than as a caricature of the opponent (or a polemical reference to some Aristotelian colleague of the time), as a splendid rhetorical move to put yourself in the other's shoes: try to imagine yourself as a Ptolemaic physicist and see what absurd consequences you will come to. The rest, net of style, is Galileo's well-known gait, rendered vividly here: the concrete examples, the stories of real experiences, the clear arguments, the extreme cases that challenge common sense. Here, you will read of cats falling from great heights without getting hurt, of vibrating chords, of theoretical digressions on the one and on infinity, of how sturdy animal bones must be, and of course of inclined planes, pendulums, projectile ranges. There are the physics of space, time, and movement, the principle of inertia, the isochronism of pendulum's oscillations, the acceleration independent of their masses in the free fall of bodies, and a lot of intelligence and beauty. But above all, thanks to De Angelis' paraphrase and algebraic translation, the genesis of Galilean ideas is better understood: not only the consolidated results, almost as if they were timeless, but the process of discovery, the concrete intellectual labor that led to their formalization. While the three friends discuss amiably, there is a world that dies, that of the traditional Renaissance academies, and a world that is emerging, the one of experience, engineering technique, the useful work of "vile mechanics."

There is still another reason to appreciate the timeliness of this work. The discourses and demonstrations you will read here owe to the lecture notes and experimental notebooks dating back to Galilei's happy Paduan period, from 1592 to 1610. Probably most of the experiments mentioned here were conceived and conducted in Padua. The characters of the Galilean narrative fiction revolve in various ways around the University of Padua and its lively intellectual environment. The book is dedicated to the Count of Noailles, for his decisive intercession in having it published (a few years after the *Dialogue concerning the two chief world systems* had been banned) in Leiden by the typographer Lodewijk Elzevir; the count had been a pupil of Galilei during his teaching period in Padua. In short, between the lines of these cordial dialogues, the University that welcomed him and gave him great freedom of research, and which in 2022 will celebrate its first eight hundred years, is omnipresent. It is therefore particularly significant that this excellent work by a scientist and professor from the University of Padua sees the light in conjunction with this impressive anniversary. From Galileo's Padua to today's Padua.

Ludovico Geymonat wrote that in the *Discourses*, together with Galileo's typical persuasive and defensive narrative, the interpenetration of mathematics and experience that will be the basis of all modern science is brought to perfection. There are epochal books, root books, and the last work of the "first mathematician and philosopher of the Grand Duke of Tuscany" is one of those, here for the first time made fully accessible to the curious readers. So it happens that a learned scientist of today, a particle physicist and astrophysicist of the twenty-first century, aware of the importance of the history of scientific ideas, manages to give us back that imperishable feeling that Galileo himself, on the fourth day of these *Discourses*, describes by writing that "the force of demonstrations such as occur only in mathematics fills with wonder and delight."

Telmo Pievani
Chair of Logics and Philosophy of Science
University of Padua, Italy

Introduction

Modern readers face major obstacles in reading Galilei, because his geometric reasoning, which irradiates culture in deference to his great Greek masters, is completely different from today's notation-dominated mathematics. This fact makes his demonstrations difficult to follow. In addition, he uses a rich and complex language, with long periods, double negations, and multiple levels of indentation. As a consequence of all of this, Galilei is extremely difficult to read and understand. understand.[1] But none of this negates the fact that he is one of the fathers of science and modern culture, as well as being quite witty and funny, and that everybody would therefore be enriched by being exposed to such artistry, such intelligence and such beauty, or, to use a more Galilean expression, by experiencing such a marvel.

This work is a translation of the *Discorsi e dimostrazioni matematiche intorno a due nuove scienze (Discourses and mathematical demonstrations related to two new sciences*, in short, *Two New Sciences*), the fundamental book by Galilei dedicated to mechanics, into modern English (largely refreshing the 1914 translation by Crew and De Salvio), and in algebraic formulae. It is written with the purpose of making Galilei comprehensible to what I imagine "modern" readers to be when I think of my children: curious, passionate about science, but, unfortunately, with little time to dig into lexical, historical, and philosophical antiquities. An operation of this kind had been carried out by Subrahmanyan Chandrasekhar with Newton's *Philosophiae Naturalis Principia Mathematica (Mathematical Principles of Natural Philosophy,* often simply called *Principia*). In contrast to Chandrasekhar's approach, I restricted myself to the mathematics available in Galilei's time (and thus, in particular, I avoided calculus), and I tried, where possible, to trace Galilei's line of thought.

---

[1]Despite its difficulty, Galilei's language, according to many Italian writers, sets a fine template for literature. Italo Calvino writes that "[...in particular when writing about the Moon] Galilei, the greatest Italian literature writer of any century, raises his prose to a degree of accuracy and evidence combined with a prodigious lyrical rarefaction. And the language of Galilei was one of the models of Leopardi, great lunar poet."

I chose to use, with a few exceptions justified in the Afterword, the original images, extracted and digitally cleaned ex novo from their initial appearance, because of their artistic nature, and because many historians (including Antonio Favaro, the editor of the national edition of the Opera Omnia of Galilei, which we refer to simply as the "national edition" in the following) attribute them to Galilei himself, who was well known for his facility in the art of drawing.[2] Thanks to the use of modern image-cleaning technologies and to the extreme care of the staff at the Biblioteca Nazionale Centrale di Firenze, I feel certain that the reproduction of the figures in this work is more faithful to the originals than in any other edition after Galilei's time.

Two New Sciences, published in 1638, was the final book released by Galileo Galilei (1564–1642). It presents a scientific work performed over the course of Galilei's life. The events described begin in 1602 and involve a long phase of meditation and discussion with numerous correspondents, Paolo Sarpi in particular, on the concepts of space, time, and movement. Galilei began writing the first draft in 1608, but in 1609, new inspiration struck: he learned of the invention of the telescope and soon developed a passion for this new instrument, subsequently improving it and becoming totally absorbed in astronomical observations for a period of several years. The writing of *Two New Sciences* became central again after 1633, following the publication of the *Dialogo sopra i due massimi sistemi del mondo (Dialogue concerning the two chief world systems)*. In addition to many original subjects, *Two New Sciences* includes topics from the *De motu*, written around 1590 and never published, and lecture notes and experimental notes dating back to the Padua period (from 1592 to 1610: "the best eighteen years of my life," according to Galilei),[3] also never published before. It represents the summa of his physical thought, just as the Dialogue concerning the two chief world systems is the summa of his cosmological thought.

The two new sciences Galilei refers to are the science of materials (related mostly to the science of construction), addressed on the first two days of discussion, and mechanics, addressed on the third and fourth days. In this work, I have chosen to include the additional day of discussion related to the strength of percussion and the origin of motion, i.e., the way movement is transmitted to a body. Galilei initially wanted to include this chapter in the first edition, as he writes twice, in the text and in a letter to the editor. By the time of publication, however, Galilei had concluded that this material was not yet sufficiently developed, and thus its release was postponed; it would not see the light of day until after his death. Having chosen to include this material, I also chose not to include an appendix on the center of gravity of solids that Galilei had composed in his youth, and that had been

---

[2]My personal opinion, also justified by the stylistic comparison with the Galilean manuscripts kept in the Italian National Library in Florence and by the progression of Galilei's blindness, is that it is likely that the figures from the first three days are attributable to Galilei, while those from the fourth day and the additional day probably are not.

[3]Letter to Fortunio Liceti, Arcetri, June 1640.

unpublished (eclipsed—as Galilei says—by Luca Valerio's *De centro gravitatis solidorum*) until it was added to the first edition of *Two New Sciences*.

Two New Sciences is one of the most important works in the history of science: it paved the way for Newton's *Principia*, published half a century later, and to experimental science in general. Newton recognized not only Galilei's authorship of the first law of dynamics, the so-called principle of inertia, but also his contribution to the second, which establishes the proportionality between force and acceleration.[4] *Two New Sciences* contains, to mention only some of its main discoveries, the principle of inertia, the description of the motion of falling bodies, the observation that bodies of different weight fall with the same acceleration in vacuo, a demonstration (correct only at the first order) of the isochronism of pendulum oscillations, a demonstration of the parabolic motion of projectiles, and innovative considerations related to acoustics and music. For the first time, physics, the science of nature, as Aristotle called it, is expressed through mathematics. For the first time, experiments are designed and performed to test hypotheses. Galilei was clearly aware of the important legacy that he was leaving, and often writes about this fact in the text. Hawking places this book among the five fundamental works in the history of physics and astronomy, and according to the mathematician Alfréd Rényi, this is the most significant mathematical work in over 2000 years.

Two New Sciences is written in the same style as the *Dialogue concerning the two chief world systems*, with the same three characters (Simplicio, Sagredo, and Salviati) engaged in discussion. Two of them were inspired by real people, friends of Galilei's: the Florentine Filippo Salviati, a member of the Accademia dei Lincei, like Galilei, and the Venetian Gianfrancesco Sagredo, formerly a pupil of Galilei's. The third character is Simplicio, a fictional character whose name is the same as that of an ancient commentator (VI Century a.D.) of Aristotle. His name implies a

---

[4]The two laws are enunciated as follows in the Principia [52]:

> Lex I: Corpus omne perseverare in statu suo quiescendi vel movendi uniformiter in directum, nisi quatenus a viribus impressis cogitur statum illum mutare

(All bodies persist in their state of rest or of uniform rectilinear motion until forces applied to them make them to change this state), and

> Lex II: Mutationem motus proportionalem esse vi motrici impressae, et fieri secundum lineam rectam qua vis illa imprimitur

(The change in motion is proportional to the motive force impressed; and is made in the direction of the straight line in which that force is impressed). Newton writes then:

> Per leges duas primas et corollaria duo prima adinvenit Galilaeus descensum gravium esse in duplicata ratione temporis, et motum projectilium fieri in Parabola, conspirante experientia, nisi quatenus motus illi per aeris resistentiam aliquantulum retardantur

(By means of these two first laws and of their corollaries Galilei found that the distance descended by heavy bodies increases with the square of time and that the motion of projectiles takes place along parabolic trajectories, as experiments confirm, neglecting some delay due to the mobile strength).

certain scientific simplicity. Simplicio often plays the role of an Aristotelian professor (a "peripatetic," as described in the text, from the name of Aristotle's school), and, as such, is not particularly critical. Sometimes, Simplicio's arguments represent the opinions of the young Galilei, Sagredo represents his middle age, and Salviati is the author in his mature age.[5] In their discussion, the three friends frequently comment upon a text written by an Academician who taught in Padua, clearly Galilei himself; they often refer to him simply as the Author, or the Academician, or sometimes even "our friend." The Author's text (quite formal, different from the dialogical part of the book) is in Latin in the original; I have transcribed it in Italics for clarity. On the additional day on the strength of percussion, which discusses the way in which movement is communicated by an impulsive force to a body, Simplicio is absent, and is replaced by Paolo Aproino from Treviso, who had been a student of Galilei's in Padua and had assisted him in some of his experiments on motion, together with Daniele Antonini from Udine. In addition to being more compelling than a treatise, as Plato had already shown and as prescribed in numerous courses on rhetoric of the time, a dialogue allows for circumventing formalism in certain demonstrations that Galileo was probably not able to develop rigorously, often due to the insufficiency of mathematics before the invention of calculus. Unlike a mathematical treatise, which follows the rule that each proposition must have been proven before moving on to the next (and the part of the book written in Latin is, indeed, a mathematical treatise), a dialogue allows its participants to forego certain rigorous demonstrations and replace them with assumptions of sufficient plausibility. In this sense, we see a role for the ambiguous meaning of the Latin word *demonstratio*, which had already been used by Cicero to signify both the act of showing, indicating and exhibiting, and the proof or formal demonstration in the mathematical sense.[6] Galilei shows here his deep knowledge of Plato's dialogues and of the tools and recipes prescribed in the classical rhetorical handbooks of Aristotle and Cicero, carefully weighing cogency, emotion and elegance as ingredients to achieve persuasion.

---

[5]In his *Dialogue...* Galilei introduces the characters as follows (from the translation by Drake): "Many years ago I was often to be found in the marvelous city of Venice, in discussions with Signore Giovanni Francesco Sagredo, a man of noble extraction and trenchant wit. From Florence came Signore Filippo Salviati, the least of whose glories were the eminence of his blood and the magnificence of his fortune. His was a sublime intellect which fed no more hungrily upon any pleasure than it did upon fine meditations. I often talked with these two of such matters in the presence of a certain Peripatetic philosopher whose greatest obstacle in apprehending the truth seemed to be the reputation he had acquired by his interpretations of Aristotle. Now, since bitter death has deprived Venice and Florence of those two great luminaries in the very meridian of their years, I have resolved to make their fame live on in these pages, so far as my poor abilities will permit, by introducing them as interlocutors in the present argument. Nor shall the good Peripatetic lack a place; because of his excessive affection toward the Commentaries of Simplicius, I have thought fit to leave him under the name of the author he so much revered, without mentioning his own. May it please those two great souls, ever venerable to my heart, to accept this public monument of my undying love. And may the memory of their eloquence assist me in delivering to posterity the promised reflections."

[6]See the Latin-Italian Dictionary by Georges and Calonghi, Rosenberg & Sellier, Turin 1950.

Since his previous book, the Dialogue, had been banned by the Church, Galilei had some trouble finding a publisher. He finally succeeded with Lodewijk Elzevir, a publisher working in Leiden, South Holland. It is most likely that the intercession of the Count of Noailles, who had been a pupil of Galilei's in his teaching period in Padua, and to whom the book is dedicated, was decisive. Elzevir wrote a beautiful preface, full of culture.

About 500 copies of the book arrived in Rome and were quickly sold. A copy reached the French mathematician Mersenne, who wrote, in the following year, a book entitled The new ideas of Galileo. Another copy reached René Descartes, who read it quickly and immediately exchanged letters with Mersenne, criticizing some of the demonstrations from the fourth day. Galilei received his author's copies only six months later, and he complained about this delay.

The word "mathematics" in the title of the book needs clarification. Although this book speaks of nature in mathematical language, Galilei, like Newton, manipulated algebraic formulas in a limited way, using geometry instead[7]: "[The Universe] is written in mathematical language, and the characters are triangles, circles, and other geometric figures, without which it is impossible to understand the world; without this, we wander around in a dark labyrinth."[8] Formulae like $F = ma$ and $E = mc^2$ are central to today's physics, but the algebraic and analytical approach was introduced by Descartes and others in the same century in which Galilei and Newton were writing their fundamental works. Galilei, like Newton in the Principia, did not use algebra, the new language: he used geometry instead, celebrating the tradition of Greek culture above the modernity of the analytical approach. The result of the complex mathematical-geometrical and literary structure of the book is, quoting Plonitsky and Reed, that "the quantity and level of mathematical argument is sufficient to dissuade many nonmathematically inclined readers from penetrating very deeply into the text. On the other hand, the text is by no means purely mathematical in nature, and the nonmathematical aspects may, in a perverse manner, dissuade the mathematically inclined from taking the text as a whole seriously enough to give it more than a selective reading. This combination, although found elsewhere in Galileo's works, presents particular complexities here, and this may help to account for the relatively low level of readership of the Two New Sciences and the prevailing, somewhat stereotyped views of the book."

Some comments for the readers. To avoid overloading the text, I used two types of notes. Those identified by a literal apex are reported at the bottom of the page, and they will not only help readers, but hopefully also amuse or amaze them; these notes are also used to indicate mistakes (according to the current physical theories) made by Galilei. Those identified with Arabic numerals are shown at the end of the book, and are quotations or comments that can be skipped at first reading. I tried to minimize the use of mathematical symbols to keep this book at a high-school level, and among the unusual symbols that I do make use of are "$\propto$" (proportional to),

---

[7]Aristotle (Metaphysics, 1025b2; On the heavens, 299a15) was convinced instead that the possibility of applying an exact science such as mathematics to real phenomena is limited.

[8]G. Galilei, Il Saggiatore.

"≡", (equal by definition to), and of the logical symbol "⇒" (implies that).
I indicate with |AB| the length of the segment AB. Finally, to make it easier for
professional readers to relate the present translation with the original work by
Galilei, references are provided at the margins of the text to the pagination of Vol.
VIII of the national edition.

To make Galilei easy to read, I have benefited from the collaboration of many
friends and colleagues, and I made some compromises; to keep this introduction
light, I will discuss all of this in the Afterword, which also describes and justifies
my stylistic choices and those related to the selection of the original material, as
well as containing a brief bibliography of previous interpretations of this book. To
those readers who, hopefully stimulated by reading this book of mine, will want to
access the thought of the Author directly in his own language, I recommend reading
the wonderful national edition by Favaro, suffused by a culture that I fear no longer
exists in the present day, but by which, fortunately, we can be enlightened thanks to
the eternity of the printed word. This book is dedicated to Antonio Favaro.

Alessandro De Angelis
Chair of Experimental Physics
University of Padua, Italy

Galilei's Units

During Galilei's life the second was an astronomical standard, but it was not a practical unit for terrestrial events. The grain of weight of the pharmacist was practically a standard throughout Europe, but it was too small for ordinary measurements. The pound, a weight unit, had different values in different countries. Length units such as the foot and the arm, in Latin called cubitus (we will call it cubit),[9] varied even more: in Italy the cubit, or braccio, indicated different lengths from one city to another, and within different centuries in the same city. Galilei did not use decimal fractions, but calculated only ratios of integers, which made small units advantageous.

Here, we report the units mostly used by Galilei for his measurements, and their translation in units commonly used today.

Space

Mile	1.65 km
Spear (lancia, picca)	3.6 m
Canna	4 cubits $\simeq$ 2.3 m
Cubit	57 cm
Foot	Half a cubit $\simeq$ 28 cm
Palm	1/3 cubit $\simeq$ 19 cm
Inch	2.5 cm
Finger	Qualitative
Punto	$\lambda \simeq 0.94$ mm

---

[9]The cubitus, already used by the Egyptians (an Egyptian cubitus corresponded to about 45 cm), is one of the oldest units of measurement of length; it is the distance between the elbow and the tip of the middle finger.

Note that the "punto" (point) is about the smallest distance that the naked eye can appreciate.

Weight

Pound	340 g
Ounce	28 g
Drachm	3 g
Denaro	1.2 g
Grain	52 mg

Time

In most of Galilei's demonstrations, an absolute measure of time was not needed; equalization of times, which could be performed using the acoustic phenomenon of beats, was enough. Using beats one can compare times with an accuracy of about 1/25 of a second. Galilei's ear, thanks also to the education given by his father, was particularly sensitive—he writes on the first day that in the fifth consonance he can feel the difference between the instants when only one of the two components is at maximum and the one when both are.

Galilei uses in the *Two New Sciences* a unit of time not very precise, the pulse-beat, and the count of oscillations of pendulums (and their beats). In his notes, he uses instead a more precise quantitative measure: the "tempo" (time), defined using a water clock more accurate than described on the third day of the *Two New Sciences*. The "tempo" corresponds roughly to the flow time of 1/30 ounce (that is, 16 grains) of water through his water clock, and it was about the smallest range appreciable with this technique.

Tempo $\tau \simeq$ 1/92 of second.

With these units, by measuring time in *tempi* and distance in *punti*, one has for the gravitational acceleration at the Earth's surface

$$g \simeq \frac{\pi^2}{8},$$

which rationalizes the relationship between length L and square of the period T ($T^2 = 4\pi^2 L/g$) in pendulums. The "tempo"'s fine calibration was likely performed in this way.

Velocity

Galilei conceived only relationships between dimensionally homogeneous quantities. Therefore, the development of his theory is, as we would say in modern terms, "manifestly covariant" with respect to the choice of units. However, he speaks explicitly of "degrees of speed," speeds reproducible by different experimenters—for example, the speed acquired by a body falling from the height of a spear.

Glossary of Some Terms Used by Galilei

The science of mechanics is today well formalized in particular thanks to the work of Newton and later of the French school of mechanics (Lagrange and Laplace to name a few). Apart from some discrepancies between Latin/Italian and English, the concepts of force, weight, energy, linear momentum (quantity of motion), momentum of a force, angular momentum, are uniquely defined and leave no room for ambiguity. The situation at the time of Galilei was very different; he used some terms (impetus, force, moment, gravity, energy, virtue, talent ...) in a way that today appears often inconsistent. It is difficult if not impossible to understand to what degree these concepts were defined in his time.

I try to explain in this short glossary what they might correspond to in the language of today's physics, analyzing the use that Galilei makes of them in various contexts of this work; already in the translation we simplified the concepts by bringing some of them together, for example, reconducting "virtue" and "talent" to other analogues.

Impetus. This term is often used in the sense of kinetic energy. In particular, Galilei says that the impetus acquired by a body falling from a height of two cubits is twice that of falling from one cubit; describing the "interrupted pendulum" experiment he says that the impetus acquired by a body descending a given height from rest is sufficient to bring it back to the initial altitude. Sometimes the word is also used as "linear momentum" (in modern terms we would say the product of mass times velocity). When talking about a definite body, it is sometimes used as a synonymous of "velocity."

Moment/Momentum (*momento*). Apart from the use as "instant of time," Galilei uses this term to indicate a combination of weight and speed (although not formalized, this corresponds to the modern *momentum*, but sometimes also to *energy*).
Finally, this term is sometimes used to indicate the momentum of a force, but in this case the distinction is clear.

Force. This term is sometimes used in the modern sense; probably this is the basis of the recognition of Newton to Galilei in his formulation of the second law in the *Principia* (see endnote 2).

Gravity. In general, it is used as "heaviness" (weight force, or weight).

Impulse. Usually, Galilei uses this term meaning "push."

Summary

In the following, to facilitate the reader, we provide a summary of the topics discussed in the dialogue. For more details see the index, where the subdivision is taken, with some small variations, from the "Table of the most notable things," an analytical index in alphabetical order that Galilei had placed at the end of his book.

First Day: First new science, concerning the resistance of solids to fracture

- Sagredo does not understand why the robustness of geometrically similar machines decreases as their size grows. Salviati states that scale counts: a horse falling from a height of 3 or 4 cubits breaks its bones while a cat falling from twice that height does not.
- The resistance of rods, columns, and ropes is discussed, and what part of this resistance can be attributed to the fear of the vacuum.[10] The case of a suspended water column, which cannot be higher than eighteen cubits, is examined as an example of resistance due to the fear of the vacuum.
- The discussion on the atomic structure of materials leads to a digression on divisibility and infinity, and to the observation that the number of squares of natural numbers is equal to the number of their roots, despite the former being a subset of the latter. In the end the conclusion is reached that if we can say that a number is infinite, this is the unit.
- The nature of light is discussed, and how its speed gives it such a large power to melt metals. Salviati describes a (failed) attempt to measure the speed of propagation of light that some observations indicate not to be instantaneous.
- Salviati states that in a medium without resistance (the vacuum) all bodies would fall at the same speed, contrary to the opinion of Aristotle who believed that velocity was proportional to weight. The discrepancies observed depend on the resistance of air. A technique to measure the weight of air is illustrated. Ebony, which has a specific weight a thousand times greater than that of air,

---

[10]According to Aristotle (Physics, IV), nature abhors empty spaces; one of the consequences is that bodies tend to remain cohesive and close to each other.

falls only a little more slowly than lead. But shape also counts: a gold leaf floats in the air and a bladder filled with air falls much more slowly than lead.

- The periods of oscillation of pendulums of the same length in vacuum are independent of the amplitude of the oscillation and of the material of which the weight is made. The relationship between length and period of a pendulum is illustrated.
- This leads to a discussion on the vibration of strings; the relation is shown between length, tension, section and specific weight of a string, and the musical note which results from its vibration. It is discussed why sometimes two musical notes heard together are consonants and sometimes dissonant.[11]

Second Day: What could be the cause of cohesion. Many of the arguments of the first day are formalized.

- Salviati demonstrates the laws regulating the balance of the lever, and introduces the concept of moment of forces. He demonstrates the laws that regulate the resistance of rods of various dimensions to fracture by the action of external forces or under their own weight.
- It is shown how the shapes of animal bones change for larger animals.
- It is shown that the optimal profile for a beam supported at one end is parabolic. It is shown how to draw a parabola, and that the curve described by a chain suspended at the two extremes is a good approximation for it.
- Hollow tubes are stronger than the full ones.

Third Day: Other new science, of local motions.

- Uniform motion is defined, and the relationship between speed, time, and distance is illustrated. Uniformly accelerated motion is then defined, and it is shown that this is the motion best describing free-falling bodies.
- It is shown that the distance traveled by a body falling from rest in a uniformly accelerated motion is proportional to the square of time. Experiments with balls rolling on inclined planes with various slopes are illustrated. It is shown that the final velocity of a body starting from rest depends only on the height of fall.
- The principle of inertia on horizontal planes is presented.
- Discussing the fall of a body along chords of a vertical circle ending in the lowest point, it is shown that the duration is the same. The problem of the determination of the curve between two points along which the descent time is minimum, the so-called brachistochrone, is discussed.
- The isochronism of pendulum oscillations is discussed.

Fourth Day: Violent motion, and the motion of projectiles.

- Conical sections and the geometrical properties of the parabola are presented.
- It shows how to compose movements along mutually perpendicular directions, by superposing them.

---

[11]Galilei, the son of a famous musician, was himself a skilled lute player.

- The motion of projectiles consists of a combination of a uniform horizontal movement and a naturally accelerated vertical movement; the result is a parabolic curve. The range of a projectile is calculated depending on the initial speed and of the initial direction. The range is maximum for an initial direction at 45° from the horizontal. The effect of the resistance of air is discussed.
- The effect of a projectile hitting a target is reduced if the target moves in the same direction as the bullet. The composition of speeds is discussed.
- A stretched string or chain can never be horizontal but describes a curve approximating a parabola.

Additional Day: The force of percussion. Simplicio is replaced by Aproino, who had been assisting Galilei in some of his experiments in Padua.

- Galilei's experiments on the force of percussion (impulsive collision—for example, striking a nail by means of a hammer, driving foundations deep underground, etc.) are presented, discussing which part in the impact can be attributed to the weight of the percutant and which part to its speed (Galilei approaches the concept of "quantity of motion" later formalized by Newton).
- The principle of inertia is presented in a more complete form with respect to the third day.

Contents

DISCORSI
E
DIMOSTRAZIONI
MATEMATICHE,

intorno à due nuoue scienze

Attenenti alla

MECANICA & i MOVIMENTI LOCALI,

del Signor

GALILEO GALILEI LINCEO,

Filosofo e Matematico primario del Serenissimo
Grand Duca di Toscana.

Con vna Appendice del centro di grauità d'alcuni Solidi.

IN LEIDA,
Appresso gli Elsevirii. M. D. C. XXXVIII.

Imprimatur

By the Bishop of Olmutz

By request of Monsignor Gio. Ernesto Platais, Bishop elected of Olmutz, I have read this treatise, in which I have found nothing that is against our holy Roman Catholic faith, or against good moral; indeed, it seems to me the illustrious and noble offspring of a happy and delicate genius, and as such, I judge that the press can give it light so that it can illuminate intelligent readers.

In the Convent of St. Michael of Olmutz of the Order of Preachers,
18 November 1636
Father Gio. Tommaso Manca de Prado, Ordinary Professor of Philosophy.

And I, Giovanni, Bishop elected of Olmutz, given that the aforementioned Reverend Father found nothing that contradicted the holy Catholic faith or the good moral, grant that the book entitled ....[a] can be printed for its use as a common good.

In Olmutz, 20 November 1636
Giovanni Ernesto, Bishop elected of the aforementioned Church.

By the Rector of the Wien University

I have studied and analyzed the book in Italian starting with the words "Largo campo...",[b] to judge that it contained nothing contrary to faith and good morals, and that therefore its printing could be authorized. And this is my opinion.

In the Caesarean Academical School of the Society of Jesus, 29 April 1637
Gualtiero Paullo s.J., Doctor in Theology and temporary Dean of the Faculty.

I consent to the printing of this book.

Wien, 29 April 1637
Leo Mylgiesser, Doctor in Medicine, University Rector.[c]

---

[a] When the work was evaluated it had not yet a title.

[b] These are the first two words of the original Italian edition.

[c] In Latin in the original.

Dedication

To the Count of Noailles
Counselor of His Christian Majesty, Knight of the Order of the Holy Spirit,
Field Marshal and Commander, Seneschal and Governor of Rouergue and   *43*
Lieutenant of His Majesty in Auvergne, my lord, and revered protector.

Most Illustrious Sir,

In the pleasure you will receive from my work I recognize your magnanimity.

You know already the disappointment and despair I felt for the unfortunate fate of my previous book; consequently, I had decided not to publish any of my works anymore. But I felt it was wise to save this manuscript from complete oblivion, at least for those who intelligently understand the problems I have faced.

Therefore I decided first of all to put my work in your hands, not being able to find a more worthy depository, hoping that, for the affection you feel for me, you could preserve my studies and my work. When you passed here returning from the Roman embassy, I wanted to greet you personally as I had done it before by letter on many occasions, and in that meeting, I presented you a copy of this work. The pleasure with which you received it convinced me that you would safeguard it. The fact that you brought it with you to France and showed it to your friends that are fond of these sciences has proven to me that my long silence would not be interpreted as idleness. *44*

Shortly after, when I was about to send a few more copies to Germany, Flanders, England, Spain, and perhaps to some places in Italy, the Elzevirs told me they were ready to print this work and I should write a dedication and send it immediately to them. This sudden and unexpected news led me to think that it was your eminence's interest in reviving and spreading my name, sending these works to several of your friends, which brought my writing into the hands of these publishers who wanted to honor me with a beautiful edition of this. But these writings must have received added value from the criticism of an excellent judge as your eminence, who won everyone's admiration for the union of many virtues. Your desire to magnify the fame of my work proves your unparalleled generosity, as well as your care for the public interest.

I have to recognize with gratitude, in a very clear way, the generosity of your eminence, which gave wings to my fame and brought it to regions far from what I could have dreamed. Therefore I must dedicate you this work of my mind. I am obliged to do so not only by the weight of the favor you have granted me, but also, if I may say so, by the interest I have in the guarantee that you will defend my reputation against opponents who could attack me.

And now, advancing under your banner and protection, I humbly wish you to be rewarded for your kindness by reaching the maximum greatness and happiness.

Arcetri, Italy
March 1638
Your most devoted servant,
Galileo Galilei

Publisher's Preface

45 Since society is held together by the services that humans provide to each other, and since arts and sciences contribute enormously to this fact, researchers in these fields have always been held in the highest esteem, and have been much appreciated by our wise ancestors. The more useful and excellent the invention is, the greater are the honor and praise received by the inventor. Sometimes inventors have been deified in order to perpetuate the memory of the good they have provided. Praise and admiration are also deserved by those privileged minds which, by limiting their attention to known things, discovered and corrected incorrect ideas in statements made by famous people and accepted for a long time as true. Although these people simply have indicated lies and did not replace them with the truth, they are still worthy of praise. And as Cicero, the prince of orators, exclaimed: "I wish it were so easy to find out the truth how to reveal lies."

The last few centuries deserve this praise because only now arts and sciences, discovered by the ancients, have been brought to ever-increasing perfection and continuously improved through investigations and experiments by clairvoyant minds; such development is particularly evident in the exact sciences. Without mentioning many other people who succeeded in this task, we must assign the first place with the unanimous approval of scholars to Galileo Galilei, a member of the
46 Accademia dei Lincei.

Galilei deserves this honor not only because he has revealed errors in many of our beliefs, as amply demonstrated in his published works, but also because through the telescope—invented in this country, but greatly improved by him—has discovered the four satellites of Jupiter, and he has shown us the true nature of the Milky Way, the sunspots, the rough and nebulous parts of the lunar surface, the triple nature of Saturn,[a] the phases of Venus and the physical nature of comets. These arguments were completely unknown to the ancient astronomers and

---

[a] Elzevir refers here to the rings of Saturn. Galilei, given the limited resolving power of his telescope, had interpreted them as two objects on the sides of the planet.

philosophers, and therefore we can say that he has revived the science of astronomy and presented it to the world in a new light.

Remembering that the wisdom, the power, and the goodness of the Creator are nowhere exhibited so well as in the heavens and in celestial bodies, we can easily recognize his great merit in bringing these bodies to our knowledge and, in spite of their almost infinite distance, in rendering them easily visible. Indeed, according to the common saying, a sight can teach more and with greater certainty in a single day than can a precept even though repeated a thousand times; or, as another says, intuitive knowledge keeps pace with accurate definition.

But the divine and natural gifts of Galilei are clear in the present work where he shows to have discovered, through many labors and vigils, two entirely new sciences, and to have demonstrated them in a conclusive, i.e., mathematical, way. What is even more remarkable in this work is the fact that one of the two sciences deals with a very old subject, perhaps the most important in nature, one which has engaged the minds of all the great philosophers and concerning which many books have been written. I refer here to motion, a phenomenon exhibiting many wonderful properties, none of which has hitherto been discovered or demonstrated by anyone. The other science which he has also developed from its very foundations deals with the resistance which solid bodies offer to fracture by external forces, a subject of great utility, especially in the sciences and in the art of construction, and also abounding in properties and theorems not hitherto observed.

This book treats for the first time these two sciences, and is full of conclusions to which, over time, others will be added by new thinkers. Moreover, through a large number of very clear demonstrations, the author paves the way for many new theorems that will be demonstrated by intelligent readers.

Day One

(First New Science, Concerning the Resistance of Solids to Fracture)

Salviati. The work that you Venetians do in your famous arsenal suggests a vast field of investigation, especially that part of the work involving mechanics. Tools and machines are built by many artisans, and among these there must be some that, partly due to inherited experience and partly due to their own observations, became very experienced and intelligent.

Sagredo. You are right. I myself, being curious by nature, frequent often the arsenal for the simple pleasure of observing the work of those who, because of their superiority to other craftsmen, are called "foremen". Discussing with them often helped me in the investigation of certain effects, including some hidden and almost unbelievable. Sometimes I fell into confusion, and I was driven to despair by not being able to explain something I observed. And I wondered if in the end the reality was simple, and if I were not like the ignorant people who complicate a problem to give the impression of knowing something about topics they don't understand.

Salv. Are you referring to that guy we asked why they used scaffoldings and rein-   *50* forcements proportionally much larger for launching a large ship than they usually do it for a small ship, and replied that they did it to avoid the danger that the ship collapses under its own weight, a danger small boats are not subject to?

Sagr. Yes, and I am referring in particular to his last statement that I always considered as a false opinion, although common: to say that these and other similar machines cannot simply be passed from small to large, because many devices working on a small scale do not work on a large scale. Now, since mechanics has its foundation in geometry, I don't see how the properties of circles, triangles, cylinders, cones, and other solid figures can change with their dimensions. If a large machine is built with all dimensions proportional to a smaller one, and if the smaller is strong enough for the purpose for which it was designed, I don't see why the bigger one shouldn't be able to withstand destructive and severe tests to which it could be subject.

© The Author(s), under exclusive license to Springer Nature Switzerland AG 2021   9
A. De Angelis, *Galileo Galilei's "Two New Sciences"*, History of Physics,
https://doi.org/10.1007/978-3-030-71952-4_1

Salv. The common opinion is in this case absolutely wrong. Machines can be built in general even more accurately on a large scale than on a small one; so, for example, a large clock can be made more precise than a small one. Some argue that a deterioration in performance in a large machine may be due to imperfections of the material; I think that the imperfections of the material are not enough to explain the differences observed between large and small machines. Even if defeats did not exist and the materials were absolutely perfect, unalterable and free from all accidental variations, and the scale were perfect, the very structure of matter causes a large machine not to be as strong as a smaller one: the larger is the machine, the larger is its weakness. Since I assume that matter is immutable, we can treat the problem on the basis of simple and pure mathematics. Thus, Sagredo, you should change the opinion that you, and perhaps many other experts in mechanics, have about the capacity of machines and structures to resist external disturbances. It is not true that if these are built with the same material and maintain the same proportions between their parts, their resistance to such external disturbances is equal or even proportional to the dimensions. We can demonstrate on a geometrical basis that a large machine is not proportionally stronger than a small one. We can also affirm that for every machine and structure, both artificial and natural, there is a limit that neither art nor nature can overcome; it is understood, of course, that the material is the same and also the proportions.

Sagr. I feel my brain in turmoil but maybe I see the light. From what you said it seems to me that it is impossible to construct two similar structures made of the same material but of different sizes, and make them proportionally strong; if so, it would be impossible to find two poles of different sizes made of the same wood with the same proportions and the same resistance.

Salv. Indeed this is what happens; and to make sure we understand each other, I'll give you an example. If we take a wooden rod of a certain thickness, fitted into a wall at right angles (i.e., parallel to the horizon), of the maximum length such that it barely sustains itself, so that, if you add a hair to its length, it breaks under its own weight, this will be the only bar of the kind in the whole world. All the longer ones will break while all the shorter ones will be strong enough to support something more than their own weight. And what I said about the ability to self-support can be applied also to other mechanical structures, so that if a beam can support the weight of ten beams equal to itself, another beam having the same proportions as the first but with different size will not be able to support ten beams equal to itself.

Please observe how the truth of facts that at first sight seem unlikely becomes clear and stands out in its naked and simple beauty. Who does not know that a horse falling from a height of three or four cubits will break its bones while a dog that falls from the same height or a cat from a height of eight or ten cubits will not suffer fractures? The fall of a grasshopper from a tower or the fall of an ant from the distance of the Moon would be equally harmless. Children fall without consequence from heights that would cost their parents a broken leg or a fractured skull. And just like smaller animals are proportionately stronger and sturdier than larger ones, so the smallest plants are more robust than the larger ones. I'm sure you both know that an oak two hundred cubits tall would not be able to support its own branches if they were distributed as in a tree of normal size, and that nature cannot produce a horse the size of twenty ordinary horses or a giant ten times higher than an ordinary man except   *53* by miracle or by a strong alteration of the proportions of the bones, which should be considerably enlarged. At the same time, the current belief that both large and small artificial machines are equally feasible and robust is an obvious mistake. A small obelisk or column can certainly be erected without danger of breakage, while the large ones will go to pieces under the minimum solicitation, or simply because of their own weight.

Here I must tell you a fact worth of attention, as are all events which occur contrary to expectations, especially when a precautionary measure is found to be the cause of a disaster. A large marble column was placed so that both its ends leaned on a piece of beam. Then somebody suggested that, to be more sure that it would not break in the middle due to its own weight, it would have been wise to pose a third support midway. This seemed an excellent idea to everyone; but then we saw that it was not good at all, since not many months passed before the column was found cracked and broken exactly above the new central support.

Simplicio. A truly remarkable and absolutely unexpected accident, especially if caused by the inclusion of that new support in the middle.

Salv. Surely this is the explanation, and when the cause is known our surprise vanishes. In fact, when the two pieces of the column were laid on the ground, it was seen that one of the terminal supports had deteriorated but that the central one had remained hard and strong, causing half of the column to remain suspended in the air without any support, and then collapse under its own weight. In these circumstances, the body behaved differently from what it would have done if supported only by external blocks: in this last case, the end would have simply followed the part that had deteriorated. This incident could not have happened to a small column, made of the same stone and with the same proportions between thickness and length.

Sagr. I trust what you tell us, but I don't understand why strength and resistance   *54* don't scale as the quantity of material; and I am the most puzzled since, on the contrary, I have noticed in other cases that strength and resistance against breaking increase with a ratio larger than the ratio of the amounts of material. For example, if two similar nails are planted into a wall, a nail that is double than the other will sustain not only twice the weight of the other, but three or four times as much.

Salv. In fact, you're not much wrong if you say eight times as much; nor does this phenomenon contradict the other, even if in appearance they seem so different.

Sagr. Salviati, why don't you give us an explanation? I imagine that this problem on resistance opens the way to many beautiful and useful ideas; if you are happy to make this topic your topic today, we will be grateful.

Salv. I would like to recall what I learned from our Academician who has thought very much on this subject, and that according to his habit has demonstrated everything using mathematical and geometrical methods, so that this could be called a new science. I can convince you with a demonstration rather than persuading you with speculative arguments, I suppose that you are familiar with mechanics as much as needed in our discussion. First of all, it is necessary to consider what happens when a piece of wood or any other solid firmly coherent is broken; this is the first principle we must know. To understand it more clearly, imagine a cylinder AB, made of wood or other solid material. We fix the upper end, A, so that the cylinder is locked vertically. At the lower end B we attach the weight C. It is clear that however great may be the tenacity and the coherence between the parts of this solid, these can be overcome by the pull of the weight C, a weight that can be increased indefinitely until the solid breaks like a rope. And as in the case of the rope whose strength we know to come from a multitude of hemp threads that compose it, so, in the case of the wood, we observe the fibers run longitudinally and make it much stronger than a single hemp rope of the same thickness. In the case of a stone or a metallic cylinder in which the coherence seems to be even greater, the cement that holds the parts together must be different from fibers and filaments; yet even these materials can be broken by a strong pull.

Simp. If it is as you say, I can well understand that the wood fibers, being as long as the piece of wood itself, make it strong and resistant against the great forces that tend to break it. But how can a rope one hundred cubits long be made with hemp fibers that are no longer than two or three cubits each, and be so robust? Furthermore, I would be happy to hear your opinion on how parts of metal, stone, and other materials that do not show a filamentous structure are put together; because, if I'm not mistaken, they show an even greater tenacity.

Salv. To solve the problems that you raise, it will be necessary to make a digression on topics that have little connection with our current purpose.

Sagr. But if by these digressions we can reach a new truth, what harm is there, since an opportunity like this, once lost, might not come back? After all, we are not tied to a fixed and short time but we meet exclusively for our amusement, and often by digressing one discovers something more interesting and beautiful than the original solution sought. I am no less curious than Simplicio and like him I want to know what is the bonding material that holds the parts of the solids together so tight that they can barely be separated.

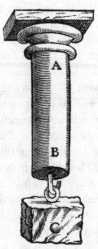

Salv. Since you want to know, I'll satisfy you. The first question is: how is it possible that fibers no longer than two or three cubits each are so closely tied together in a rope of a hundred cubits that a great force is necessary to break it? Now tell me, Simplicio, can't you hold a hemp fiber so tightly between your fingers that I, pulling by the other end, would break it before drawing it away from you? Of course you can! And now, when hemp fibers are retained not only at the extremities, but they are grasped by the surrounding medium for all their length, isn't it harder to loosen them from what holds them than to break them? But in the case of the rope the fact that the threads are twisted causes them to tie together in such a way that when the string is stretched with great force the fibers break rather than separate from each other. At the point where a rope breaks the fibers are very short, not even a cubit long, as they would be if they slipped over each other.

Sagr. To confirm this it may be remarked that ropes sometimes break not by a lengthwise pull but by excessive twisting. This, it seems to me, is a conclusive argument because the threads are bound one another so tightly that the compressing fibers do not permit those which are compressed to flow and lengthen even by the little that would be necessary to avoid fracture.

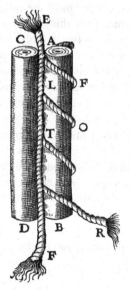

Salv. You are right. Now you see how a problem may shed light on another. The thread held between the fingers does not yield to one who wishes to draw it away even when pulled with considerable force, but resists because it is held back by a double compression, seeing that the upper finger presses against the lower as hard as the lower against the upper. Now, if we could only maintain one of these pressures, there is no doubt that only half of the original resistance would remain; but since we are not able, lifting, for example, the upper finger, to remove one of these pressures without also removing the other, it becomes necessary to preserve one by means of a new device that causes the thread to press against the finger or against some other solid body on which it rests; and so it is realized that the same force that pulls it to tear it away compresses it more and more as traction increases. This can be obtained, for example, by winding the thread around a cylinder in the form of a spiral. I explain myself better by a figure. Let AB and CD be two cylinders between which the thread EF is stretched; to be more clear we will imagine it to be a small cord. If these two cylinders be pressed strongly together, the cord EF, when drawn by the end F, will undoubtedly stand a considerable pull before it slips between the two compressing solids. But if we remove one of these cylinders the cord, though remaining in contact with the other, will not thereby be prevented from slipping freely. On the other hand, if one holds the cord loosely against the top of the cylinder A, winds it in the spiral form AFLOTR, and then pulls it by the end R, it is evident that the cord will begin to bind the cylinder; the greater the number of spirals the more tightly will the cord be pressed against the cylinder by any given pull. As the number of turns increases, the contact line becomes longer and consequently more resistant, and so the cord always resists more and more to the traction force. It is not this is precisely the type of resistance encountered in the case of a thick hemp rope whose fibers form thousands and thousands of similar turns? The effect of these weaves is so great that few short threads form the strongest ropes, which I think are called *suste*.

Sagr. What you say reminds me of two things I can't understand. The first is how it is possible that a cord that only makes two or three turns around a drum, and that is held to a headed by a slender boy, can not only hold a heavy load, but also lift it if the boy begins to rewind it. The second is linked to an astute way that a young relative of mine invented to get out of the window with a rope, avoiding hurting his palms, as it had happened to him in his first attempts. To facilitate understanding, I draw a little sketch. My relative used a wood cylinder AB about the same thickness as a walking stick and about a foot long. On this cylinder he engraved a groove in the form of a spire, not more than a wrap and a half, of sufficient width to slide the cord he wanted to use. He inserted the cord from the head A and, after having made it pass inside the furrow, made it exit from B. Subsequently he covered both the cylinder and the cord with a hinged tin tube from the side, so that it could open and close it. He secured the cord at the top to a secure support and he grabbed the tube with both hands so that he was suspended with his arms. The friction between the cord and the cylinder was such as to allow not to fall and decide, loosening the cord, at what speed descend.

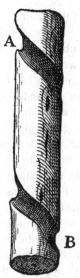

Salv. A truly ingenious device! However, I feel that for a complete explanation we   *59*
must take into account further considerations, but I do not want to digress on other
topics while you wait to hear my opinion about the breaking resistance of materials
that, unlike of the rope and most of the woods, do not possess a filamentous structure.
Cohesion of these bodies is due, in my opinion, to two causes: the first is that nature
fears emptiness; this is not enough, however, and a binder or glue or a viscous
substance is needed to join the particles that make up the body.

I will speak first of the vacuum,[a] identifying with appropriate experiments the
nature and the magnitude of its force.

If you take two well polished and smooth plates of marble, metal, or glass and
place them face to face, one will easily slide over the other, showing that there is
no glue joining them. But when you attempt to separate them, they exhibit such a
repugnance to separation that the upper one will carry the lower one with it and keep
it lifted indefinitely, even when the latter is big and heavy.

If instead the two plates had not been carefully cleaned, and consequently the
contact is not uniform, when you attempt to separate them slowly the only resistance
offered is that of weight; if, however, the pull is sudden, then the lower plate rises, but
quickly falls back, having followed the upper plate only for that very short interval

---

[a] In this book we use, in accordance with Crew and De Salvio [11], the word "vacuum" to identify
the empty space. Others (notably Drake [10]) use instead the word "void".

The dispute over whether vacuum can exist has been long, and its first records date to the Greek
atomism—vacuum (χενόν) and atoms were the fundamental constituents of matter. Plato believed
that, to be perceptible by senses, the vacuum needed some properties incompatible to its definition.
Aristotle believed that no vacuum could occur naturally, because the denser surrounding material
continuum would immediately fill it: this is the "fear of the vacuum", which causes cohesion.
Aristotle allowed however the existence of vacuum as a "filler" inside matter.

The discussion was still lively at Galilei's times, and it will soon become central between Pascal
and Descartes. It will become central, again, in 20th century physics.

of time required for the expansion of the small amount of air remaining between the plates, in consequence of their not fitting, and for the entrance of the surrounding air.

This resistance between the two plates is likewise present between the parts of a solid, and enters, at least in part, as a concomitant cause of their coherence.

Sagr. Please let me express a consideration that just came to my mind. Seeing the lower slab that follows the upper one and with a very fast movement is raised, one could say that motion in the vacuum is not instantaneous, thus refuting the theses of many philosophers including Aristotle. In fact, if it were instantaneous, the two plates would immediately separate without resistance in that moment of time necessary for the air coming from outside to fill the vacuum.[1] In following what the lower slab does on the upper one understands how motion is not instantaneous in the vacuum; it is also clear that an empty space remains between the two slabs, even if only for very little time, equal to that necessary for the surrounding medium to fill the vacuum, because if there were no vacuum there would be no need of any motion in the medium. One must then admit that a vacuum is sometimes produced by violent motion contrary to the laws of nature (although in my opinion nothing occurs contrary to nature except the impossible, that never occurs).

But here I find another difficulty: although the experiment convinces me of the correctness of this conclusion, my mind is not entirely satisfied as to the cause to which this effect is to be attributed. The separation of the plates precedes the formation of the vacuum which is produced as a consequence of this separation; and since it appears to me that, in the order of nature, the cause must precede the effect, even though it appears to follow in point of time, and since every positive effect must have a positive cause, I do not see how the adhesion of two plates and their resistance to separation can be referred to vacuum as a cause when this vacuum must still be produced. According to the infallible opinion of the Philosopher, the nonexistent cannot produce any effect.[2]

Simp. Since you grant this axiom to Aristotle, I don't think you can deny another which is excellent and reliable: nature does not begin to do what cannot be done.[3] And from this axiom it seems to me that you can solve your doubt. Nature does not conceive the vacuum nor what derives from it, and therefore does not allow the separation between the two plates.

Sagr. Assuming that the solution to my doubt is this one proposed by Simplicio, it seems to me that this same resistance to the vacuum should be enough to hold together the parts of a solid like those of a stone or of a metal or the parts of any other solid which is knit together more strongly and which is more resistant to separation. If for one effect there is only one cause, as I believe, or if, when there are more causes, they can be reduced to one, why cannot be this fear of vacuum, which indeed exists, a sufficient cause for all kinds of resistance?

Salv. For now, I do not want to participate in this discussion on whether the vacuum can hold the disjointed parts of the solid bodies together, but I tell you that the fact that the vacuum holds the two plates together is not enough to explain the connection of the parts of a cylinder of marble or metal, which, if stimulated by strong forces that pull them directly, eventually separate and divide. And when I found a way to distinguish this already known resistance, which depends on the vacuum, from any other resistance, I understood that any resistance contributes to strengthening the cohesion of the particles. If I show how the fear of the vacuum is not enough to explain this effect, wouldn't you think that it is necessary to introduce another cause?

Simp. It is clear that Sagredo's doubts must be related to something else than a so clear and logical demonstration.

Sagr. I think you're right, Simplicio. I thought that, given that a million in gold every year is not enough for Spain to pay its soldiers, it is necessary to add something else taken by many small local taxes. But go ahead, Salviati, and supposing I accept what you say, show us how fear of the vacuum is not enough to explain the effect we are talking about, and how to separate what is due to the fear of the vacuum from other causes.

Salv. I will then tell you how to separate the force of the vacuum from the other causes and how to measure it. To this end let us examine a continuous substance whose parts have as their only resistance the fear of the vacuum, water for example, as demonstrated by our Academician in one of his treatises.[4] Whenever a cylinder   *62* of water is subjected to a pull and offers a resistance to the separation of its parts, this can be attributed to no other cause than the resistance of the vacuum.

To make this measurement I have invented a device that I can explain by means of a sketch.

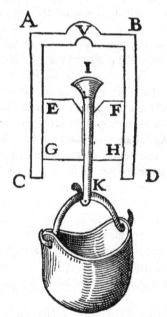

Let CABD represent the cross section of a cylinder made of metal or, preferably, of glass, hollow inside, and well turned. Let us introduce into this cylinder, two or three fingers far from the end, a perfectly fitting inner cylinder of wood EGFH, capable of moving up and down. A hole is bored through the middle of this cylinder to receive an iron wire, carrying a hook at the end K, while the upper end of the wire, I, is provided with a conical head. The wooden cylinder is cut at the top in order to receive with a perfect fit the conical head I of the wire IK, when pulled down.

Let us fill the two or three fingers left between EF and AB with water, by holding the vessel with the mouth CD upwards, pushing down on the stopper EFGH, and at the same time keeping the conical head of the wire, I, away from the hollow portion of the wooden cylinder. The air is thus allowed to escape alongside the iron wire as soon as one presses down on the wooden stopper. The air having been allowed to escape and the iron wire having been drawn back so that it fits against the conical depression in the wood, invert the vessel, bringing its mouth downwards, and hang on the hook K a vessel which can be filled with sand or any heavy material in quantity sufficient to finally separate the upper surface of the stopper, EF, from the lower surface of the water to which it was attached only by the resistance of the vacuum. Then weigh the stopper and the wire together with the attached vessel and its contents; the total weight will measure the force of the vacuum.

63

If one attaches to a cylinder of marble or glass a weight which, together with the weight of the marble or glass itself, is just equal to the sum of the weights before mentioned, and if breaking occurs, we shall then be justified in saying that the vacuum alone holds the parts of the marble and glass together; but if this weight does not suffice and if breaking occurs only after adding, say, four times this weight, we shall then be compelled to say that the vacuum provides only one fifth of the total resistance.

Simp. This invention is clever, but I have found many difficulties. Is it certain that air cannot penetrate between the glass and the cylinder? In order for the cone to fit perfectly, maybe it may not be enough to grease it with wax or resin. Moreover, couldn't water expand and dilate? Is it excluded that water, air or some exhalations penetrate into the porosity of the wood or into the same glass?

Salv. With great skill you are showing us the problems that could arise, and in part giving us the remedies to prevent air from penetrating the wood or passing between the wood and the glass. But now let me point out that, as our experience increases, we shall learn whether or not these potential difficulties really exist. If, as is the case with air, water is by nature expansible, although only under severe treatment, we shall see the stopper descend; and if we put a small excavation in the upper part of the glass vessel, indicated by V, then the air or any other tenuous and gaseous substance, which might penetrate the pores of glass or wood, would pass through the water and collect in this receptacle V. But if these things do not happen we are guaranteed that our experiment has been performed with proper caution, and we shall discover that water does not dilate and that glass does not allow any material, however tenuous, to penetrate it.

Sagr. Thanks to this discussion, I have learned the cause of a certain effect which I have long tried to understand. I once saw a cistern that had been provided with a pump under the wrong assumption that water might thus be drawn with less effort or in greater quantity than by means of the ordinary bucket. The stock of the pump carried its sucker and valve in the upper part so that the water was lifted by attraction and not by a push as is the case with pumps in which the sucker is placed lower

64   down.

This pump worked perfectly as long as the water in the cistern stood above a certain level, but below this level the pump failed to work. When I first noticed this phenomenon I thought the machine was out of order, but the workman whom I called in to repair it told me the defect was not in the pump but in the water which had fallen too low to be raised through such a height; and he added that it was not possible, either by a pump or by any other machine working on the principle of attraction, to lift water a hair's breadth above eighteen cubits; whether the pump is large or small this is the extreme limit of the lift.[b]

I knew that a rope, or rod of wood, or of iron, if sufficiently long, would break by their own weight when held by the upper end, but I had never thought before that the same thing would happen, only much more easily, to a column of water. And a pump consists just of a column of water attached at the upper end; if you stretch it more and more you will finally reach a point is where it breaks, like a rope, due to its weight.

Salv. This is the way it works. And since the height of eighteen cubits is the maximum possible for any amount of water, whatever the pump, large, small, or even as thin as a straw, we can say that by weighing the water contained in a tube eighteen cubits long, regardless of the diameter, we will obtain the value of the force of the vacuum in a cylinder of any solid material having the same diameter. It is easy to find to what length cylinders of metal, stone, wood, glass, etc., can be stretched without breaking under their own weight.

65

Take for example a copper wire of any length and thickness; fix the upper end and at the other attack a load larger and larger, which eventually will lead the wire to break; let the maximum load be, say, 50 pounds. Then it is clear that if 50 pounds of copper, in addition to the weight of the wire itself which might be, say, 1/8 ounce, is drawn out into wire of this same size, we shall have the greatest length of this kind of wire which can sustain its own weight.

Suppose the wire that breaks to be one cubit in length and 1/8 ounce in weight; then since it supports 50 pounds in addition to its own weight, i.e., 4800 eighths of an ounce, it follows that all copper wires, independent of size, can sustain themselves up to a length of 4801 cubits and no more. The part of the breaking strength which depends upon the vacuum is equal to the weight of a rod of water, eighteen cubits long and as thick as the copper rod. If, for example, copper is nine times as heavy as water, the breaking resistance of any copper rod, depending upon the vacuum, is equal to the weight of two cubits of this same rod (the length is inversely proportional to the specific weight).

With a similar method we can find the maximum length of a wire or rod of any material which can support its own weight, and discover the part that the strength caused by the force of vacuum on the resistance to fracture.

Sagr. It remains to be understood what the resistance to fracture depends on, other than the resistance of vacuum; what is the gluey or viscous substance which cements

---

[b] Galilei gives a naive interpretation of the phenomenon, which was explained by his disciple Evangelista Torricelli (1608–1647). It is not the pump lifting the water, but atmospheric pressure pushing it, and this pressure corresponds to that of a column of water about 10.3 m (18 cubits) high.

together the parts of the solid? I cannot imagine a glue that will not burn up in a highly heated furnace in two or three months, or certainly within ten or a hundred. And yet if gold, silver, and glass are kept melted for a long time and are removed from the oven, their parts, upon cooling, come together immediately and tie together as before, and the same for glass and cement. What holds these parts together so firmly?

66

Salv. Given the evidence that the fear of vacuum prevents the separation of two plates except with violent effort and the same can be said of two large parts of a marble or bronze column, I cannot see why this repugnance for emptiness cannot be in the same way the cause of the coherence between the very smallest particles of these materials. Since each effect must have one true and sufficient cause and since I find no other cement, am I not justified in trying to discover whether the vacuum is not a sufficient cause?

Simp. If you have already shown that the resistance to the separation of two large surfaces due to vacuum is very small compared to the one that holds small particles together, don't think that you demonstrated that there are at least two different causes at work?

Salv. Sagredo has already answered this question when he remarked that each individual Spanish soldier was paid with money collected from many small sums, given that a million in gold would not be enough to pay the entire army. And who knows if many small voids don't operate between the small particles?

Now I'll tell you something which has just occurred to me and which I do not offer as an absolute fact, but rather as a passing thought, still immature and calling for more careful consideration.

Sometimes I have observed how fire winds its way in between the most minute particles of this or that metal and, even though these are solidly cemented together, separates them. After removing fire, these particles reunite with the same tenacity as before, without any loss of quantity in the case of gold and with little loss in the case of other metals, even though these parts have been separated for a long while, I have thought that the explanation might lie in the fact that the extremely fine particles of fire, penetrating the pores of the metal (too small to admit even the finest particles of air or of many other fluids), would fill the small intervening vacua and would set free these small particles from the attraction which these same vacua exert upon them

67

and which prevents their separation.

Thus particles are able to move freely so that the material becomes fluid and remains so as long as the particles of fire remain inside. If they leave the vacua, the original attraction returns, and the parts are again cemented together.

And I would reply to Simplicio's observation that, although such vacua seem very small, their large number multiplies the resistance. The nature and quantity of strengths that result from adding an immense number of small forces are clearly illustrated by the fact that a weight of millions of pounds, suspended by large cables, is exceeded and relieved when the southern wind carries innumerable atoms of water, suspended in a fine mist, which moves through the air and penetrates between the fibers of the ropes, stretching them despite the tremendous force of the suspended weight. When particles penetrate the small pores of the ropes they reduce their size so that a very heavy load can be lifted.[c]

Sagr. There is no doubt that any resistance, as long as it is not infinite, may be overcome by a multitude of minute forces. A vast number of ants might carry a ship full of grain: one ant can easily carry one grain, and since the number of grains in the ship is not infinite, if you take a number four or six times as great of ants they will carry the grain ashore and the boat as well. This will indeed call for an enormous number of ants, but in my opinion this is precisely the case with the vacua which bind together the smallest particles of a metal.

Salv. And if this demanded an infinite number of particles, would you think it impossible?

Sagr. Not if the mass of metal were infinite, otherwise...

Salv. Salv. Otherwise what? Since we have arrived at paradoxes let us see if we cannot prove that within a finite extent it is possible to discover an infinite number of vacua. At the same time, we shall solve the most remarkable of the list of problems that Aristotle himself in his *Mechanical Problems*[5] calls marvelous.[d] This solution might be as conclusive as the one he himself reaches, and different from the one given by the most learned Lord Bishop de Guevara.[6]

First it is necessary to consider a proposition, not discussed by others, but upon which the solution to the problem depends, and from which, if I am not mistaken, we shall derive new and remarkable conclusions. For the sake of clearness let us draw a figure.

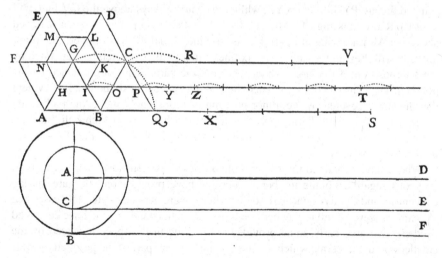

---

[c] It seems that during the elevation of the Vatican obelisk of St. Peter's Square in 1586 the ropes supporting the monument were put under tension by moistening them. The story was well known in the time of Galilei.

[d] Galilei often uses this word (in Italian meraviglioso). It means that it has the effect of arousing wonder, marvel and astonishment.

Consider an equilateral and equiangular polygon, with any number of sides, centered at point G, for example a hexagon ABCDEF, and draw another one similar to it, concentric but smaller, that we call HIKLMN.

We prolong the side AB of the major hexagon towards S, and in the same way the corresponding side HI of the minor hexagon in the same direction, so that the HT line is parallel to AS; and let another line GV pass through the center. Once this is done, let's imagine that the larger polygon rotates on the AS line, taking with it the minor polygon. It is evident that when point B, at the end of the side AB, remains at the beginning of the rotation, point A rises, and point C falls by describing the arc CQ until the BC side coincides with the line BQ, equal to BC. But during this rotation, the point I of the smaller polygon rises above the IT line, and returns to the IT line when point C has reached position Q.

The point I, having described the arc IO above the HT line, will reach the position O at the same time in which the IK side has taken the OP position; but in the meantime the G center has left the GV line and not will come back to it until the GC arc is completed. After this, the major polygon was brought to stand with its side BC on the BQ line, and the IK side of the minor polygon has been made to coincide with the OP line, having passed the entire IO part without touching it; also the G center will have reached position C after having crossed its entire course over the parallel GV line. And finally, the whole figure will assume a position similar to the first; so if we continue the rotation and get to the second step, the side of the major polygon CD will coincide with the QX portion and the KL side of the minor polygon, having first skipped the arc PY, falling on YZ, while the center, always above the GV line, will return to R after missing CR. After a full rotation, the larger polygon will have traced above the AS line, without interruptions, six lines equal to its perimeter; the minor polygon will draw in the same way six lines equal to its perimeter, but separated by the interposition of five arcs, which represent the parts of HT not touched by the polygon. The center G does not reach the GV line, except in six points. It is evident that the space covered by the minor polygon is almost the same as the path taken by the major, i.e., the HT line approximates the AS line, from which it differs only in length of a chord of one of these arcs, considering however that the HT line also includes the five arcs.

What I have explained by the example of these hexagons also happens with all polygons, regardless of the number of sides they have, provided only they are similar, concentric, and rigidly connected, so that when the greater one rotates, the lesser will also turn however small it may be. You must also understand that the lines described by these two are nearly equal provided we include in the space traversed by the smaller one the intervals which are not touched by any part of the perimeter of this smaller polygon.

A polygon of a thousand sides, having completed a rotation, draws a line equal to its perimeter; at the same time, a minor polygon similar to the larger one describes a line approximately equal, but composed of small portions equal to its thousand sides with the interposition of a thousand spaces that, compared to the lines that correspond to the sides of the polygon, we can call vacua. So far there is neither difficulty nor doubt.

But now suppose that about any center, say A, we draw two concentric and rigidly connected circles; and suppose that from the points C and B, on their radii, we draw the tangents CE and BF and that through the center A the line AD is drawn parallel to them. If the large circle makes one complete rotation along the line BF, equal not only to its circumference but also to the other two lines CE and AD, tell me what the smaller circle will do and also what the center will do. The center will certainly traverse and touch the entire line AD while the circumference of the smaller circle will touch the entire line CE, just as was done by the above mentioned polygons. The only difference is that the line HT was not at every point in contact with the perimeter of the smaller polygon, but there were as many vacant spaces left untouched as there were spaces coinciding with the sides. But here in the case of the circles the circumference of the smaller one never leaves the line CE, so that no part of the latter is left untouched, nor is there ever a time when some point on the circle is not in contact with the straight line. How now can the smaller circle traverse a length greater than its circumference unless it proceeds by jumps?

Sagr. I was thinking if one may say that, just as the center of the circle, carried along the line AD, is constantly in contact with it, although it is only a single point, so the points on the circumference of the smaller circle, carried along by the motion of the larger circle, would slide over some small parts of the line CE.

Salv. This cannot be true. First I see no way that a point of contact, such as that   *71* at C, could slip over certain portions of the line CE. But if such slidings along CE did occur they would be infinite in number since the points of contact are infinite in number: an infinite number of finite slips will however make an infinitely long line, while the line CE is finite. The other reason is that as the greater circle, in its rotation, changes its point of contact continuously, the lesser circle must do the same because B is the only point from which a straight line can be drawn to A and pass through C. Accordingly the small circle must change its point of contact whenever the large one changes: no point of the small circle touches the straight line CE in more than one point. Not only so, but even in the rotation of the polygons there was no point on the perimeter of the smaller which coincided with more than one point on the line traversed by that perimeter. This is at once clear when you remember that the line IK is parallel to BC and that therefore IK will remain above IP until BC coincides with BQ, and that IK will not lie upon IP except at the very moment when BC occupies the position BQ; at this instant, the entire line IK coincides with OP and immediately afterward rises above it.

Sagr. The question looks very complicated. Please explain to us better.

Salv. Let's return to the consideration of the polygons mentioned above and of which we already understood the behavior. In the case of polygons with 100000 sides, the line traversed by the 100000 sides of the perimeter of the major is equal to the line drawn by the 100000 sides of the smaller, provided we include the 100000 interleaved empty spaces. In the case of circles, which are polygons having an infinite number of sides, the line crossed by the infinite sides continuously distributed in the greater circle is equal to the line drawn by the infinite sides in the smaller circle, except that the latter alternate with empty spaces; and since the sides are not finite in number, but infinite, so even the empty spaces are infinite. The line crossed by the larger circle consists then of an infinite number of points that fill it completely, while the path traced by the circumference of the smaller circle consists of an infinite number of points that leave empty spaces and only partly fill the line.

After dividing and resolving a line into a finite number of parts, that is, into a number which can be counted, it is not possible to arrange them again into a greater length than that which they occupied when they formed a continuum and were connected without the interposition of as many empty spaces. But if we consider the line resolved into an infinite number of infinitely small and indivisible parts, we shall be able to conceive the line extended indefinitely by the interposition of an infinite number of infinitely small indivisible empty spaces.

Now, what has been said about simple lines must also be true in the case of surfaces and solid bodies, noting that they consist of an infinite number of atoms. A body, once divided into a finite number of parts, is impossible to reassemble so as to make it occupy more space than before, unless a finite number is interposed of empty spaces, i.e., spaces free from the substance of which the solid is made. But if we imagine a body made of an infinite number of fundamental elements, then we will be able to think of them as extended indefinitely in space, with the interposition of an infinite number of empty spaces. So you can easily imagine a gold ball expanded to a very large space with the introduction of an infinite number of empty spaces, provided that gold is composed of an infinite number of fundamental parts.

Simp. It seems to me that you are moving towards the theory of scattered voids, supported by Democritus.

Salv. But you did not add, "denying Divine Providence," an inapt remark made on a similar occasion by a certain antagonist of our Academician.[e]

Simp. I noticed, and not without indignation, the rancor of this bad opponent; further references to these affairs I omit, not only as a matter of good form, but also because I know how unpleasant they are to the good tempered and well ordered mind of somebody so religious and pious, so orthodox and God-fearing as you. But returning to our subject, your previous explanation leaves with me many difficulties which I am unable to solve. First, if the circumferences of the two circles are equal to the two straight lines, CE and BF, the latter considered as a continuum, the former as interrupted with an infinity of empty points, I don't see how it is possible to say that the line AD described by the center, and made up of an infinity of points, is equal to this center which is a single point. Besides, this building up of lines out of points, divisibles out of indivisibles, and finite out of infinite, offers me an obstacle

[e] Based on a note written by Galilei (Tome III of the Astronomical Works, Florence, 1843), could be the Jesuit Orazio Grassi (1583–1654). Grassi replied with the treatise *Astronomical libra ...* (1619), under the pseudonym Lotharius Sarsius (an anagram of Horatius Grassius), to Galilei's opinion that comets were made of steam that lights up when heated by the Sun. Galilei in 1623 replied to the *Astronomical libra ...* with the work *Il Saggiatore* (*The Assayer*; the assayer is a precision balance, as opposed to the *libra* which was the common balance), in which he reaffirmed his theory, also suggesting that he appreciated atomism, and speaking of the corpuscular nature of light. *Il Saggiatore* was received with great favor in the circles of the curia; Grassi, accused of anger and envy, is said to have filed an anonymous complaint against Galilei accusing him to deny Divine Providence with his atomistic theses.

difficult to avoid; and the necessity of introducing vacuum, so conclusively refuted by Aristotle, presents the same difficulty.

Salv. There are indeed difficulties, but let's remember that we are talking about infinities and indivisibles, incomprehensible to our intellect, the former for their magnitude, the latter for their smallness. We, humans, cannot however avoid facing them; so, I will take the liberty to fantasize with ideas that are certainly not conclusive, but at least fascinating for their novelty. But maybe this is moving us away from the initial problem and it may appear to you inappropriate and unwelcome.   *73*

Sagr. We are grateful to enjoy the benefit and privilege of speaking between friends about freely chosen and futile things, unlike dealing with books of the past, which generate a thousand doubts without solving any. Let me share the considerations that came up during your reasoning, because we have time, thanks to the fact that we are not obliged to perform necessary tasks, to continue to solve the other problems that emerged and, in particular, the doubts raised by Simplicio, which should not be neglected.

Salv. Granted. The first question was: how can a point be equal to a line? Since I cannot do more for now, I shall attempt to remove, or at least diminish, one improbability by introducing a similar or a greater one, just as sometimes wonder is diminished by a miracle. And this I shall do by showing you two equal surfaces, together with two equal solids located upon these same surfaces as bases, all four of which diminish continuously and uniformly in such a way that their remainders always preserve equality among themselves. Finally, both the surfaces and the solids terminate their previous constant equality by degenerating, the one solid and the one surface into a very long line, the other solid and the other surface into a single point; that is, the latter to one point, the former to an infinite number of points.

Sagr. This seems to me a marvelous concept, but let's hear the explanation and   *74* the demonstration.

Salv. I need to make a drawing.

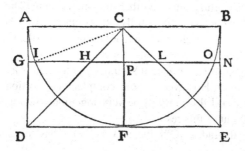

Let AFB be a semicircle with center C. Let us draw around it the ADEB rectangle. From the center we draw the straight lines CD and CE up to the points D and E respectively. We rotate the whole figure around the axis of rotation CF, perpendicular to AB and DE. It is clear that the rectangle ADEB is the section of a cylinder, the semicircle AFB of a hemisphere, and the triangle CDE of a cone. Next let us remove the hemisphere but leave the cone and the rest of the cylinder, which, on account of its shape, we will call a "bowl." First, we shall prove that the bowl and the cone are

equal in volume; then we shall show that a plane parallel to the circle which forms
the base of the bowl and which has the line DE for diameter and F for a center—a
plane whose trace is GN—cuts the bowl in the points G, I, O, N, and the cone in
the points H, L, so that the part of the cone indicated by CHL is always equal to the
part of the bowl whose profile is represented by the triangles GAI and BON. Besides
this, we shall prove that the base of the cone, i.e., the circle whose diameter is HL,
is equal to the circular surface which forms the base of this portion of the bowl, or
as one might say, equal to a ribbon whose width is GI. Note by the way the nature
of mathematical definitions which consist merely in the imposition of names or, if
you prefer, abbreviations introduced in order to avoid a tedious work. Indeed we can
call this surface a "circular band" and that sharp solid portion of the bowl a "round
75   razor."

The conclusion seems a miracle: as the cutting plane moving upwards approaches
the AB line, the portions of the solids that are cut are always equal in volume, as
well as the areas of theirs bases. And as the cutting plane approaches the top, the
two solids (always of equal volume) as well as their bases (which always have equal
areas) tend to disappear, but in the first case they end in the circumference of a circle,
that is the upper edge of the bowl, and in the second in a single point, that is the apex
of the cone. Since, as these solids are reduced, they are equal until the end, we are
justified in saying that at the end of this reduction they are still the same, and one is
not infinitely greater than the other. So it seems we can equate the circumference of
a large circle to a single point. And what is true of solids is also true for the surfaces
that form their bases, because these also maintain equality between them through
their diminishing and eventually vanishing, one in the circumference of the circle,
the other in a single point. Why should not then call them equal, since they are the last
elements of a series of elements of equal size? Note that even if these volumes were
large enough to contain the celestial hemispheres, the upper edge of the bowl and the
apex of the cone would always remain equal and would vanish, one in the maximum
circles of the celestial orbits, and the other in a single point. Then in analogy with
what has been said, we can say that all the circumferences of the circles, however
different, are equivalent to each other, and are in turn equivalent to a single point.[7]

Sagr. This presentation seems so intelligent to me that, even if I were able, I
wouldn't want to oppose it, because it would seem to me a sacrilege to ruin such
76   a beautiful construction. However, for our complete satisfaction, please give us a
demonstration, because I think it will be very ingenious, seeing how subtle is the
philosophical argument of this result.

Salv. If we call r the radius $|CP|$, the area A_C of the base of the cone (a circle) is

$$A_C = \pi r^2 .$$

The area of the circular band at the base of the bowl is

$$A_S = \pi(|PG|^2 - |PI|^2)$$

(difference between the areas of the two circles of radii PG and PI), but since $|PG| = |CI|$, and $(|CI|^2 - |PI|^2) = |CP|^2 = r^2$, we have

$$A_S = \pi(|PG|^2 - |PI|^2) = \pi r^2 = A_C \, ,$$

as to be demonstrated.

Now, since the cone of radius r and the corresponding "round razor" are composed of equal surfaces (the sections of a generic plane between 0 and r) and have the same height, their volumes are equal.[8]

Sagr. The demonstration is ingenious and its consequences are remarkable. Now let's hear something concerning the other difficulty raised by Simplicio.

Salv. Infinity and indivisibility are in their very nature incomprehensible to us; imagine then what they are when combined! If we wish to build up a line out of indivisible points, we must take an infinite number of them, and are, therefore, bound to understand both the infinite and the indivisible at the same time. Many ideas have passed through my mind concerning this subject, some of which, possibly the most important, I may not be able to recall now; but in the course of our discussion it may happen that I shall awaken in you, and especially in Simplicio, objections and difficulties which in turn will bring to memory that which, without such stimulus, would have been dormant in my mind. Allow me therefore the customary liberty of introducing some of our human fancies, as we may call them in comparison with supernatural truth which provides the one true and safe recourse for decision in our discussions and which is an infallible guide in the dark and dubious paths of thought.[f]

One of the main objections against this building up of continuous quantities out of indivisible quantities is that the addition of one indivisible to another cannot produce a divisible: if this were so, it would render the indivisible divisible. Thus if two indivisibles, say two points, can be united to form a quantity, say a divisible line, then an even more divisible line might be formed by the union of three, five, seven, or any other odd number of points. Since however these lines can be cut into two equal parts, it becomes possible to cut the indivisible which lies exactly in the middle of the line. In answer to this and other objections of the same type we reply that a divisible cannot be c constructed out of two or ten or a hundred or a thousand indivisibles, but requires an infinite number of them.

Simp. I have a problem that seems to me impossible to solve: given that a line can be greater than another, and both contain an infinite number of points, we must admit that within a line we have something larger than the infinite because the infinite points of the major line are more than the infinite points of the minor line. This means assigning to an infinite quantity a value greater than the infinite. This is a concept that I cannot understand.

---

[f] Here Galilei is very cautious because the idea that a line could be composed of indivisible, hypothesized by Epicurus but strongly opposed by Aristotle, had been condemned as a heretic in 1415 by the Council of Constance. The corpse of John Wyclif (1331–1384), a theologian and teacher in Oxford, had been exhumed and burned in 1428 as punishment for this and other Epicurean doctrines.

Salv. These are some of the difficulties arising when, with our finite minds, we try to discuss the infinite by assigning to it attributes that we would give to finite and determined things.

But this is wrong, because we cannot speak of infinite quantities as if they were greater than, less than, or equal to others. In order to prove this, I thought of an argument to make it clear put in the form of a question to Simplicio, who raised this doubt. I take for granted that you know which integer numbers are squares and which are not.

Simp. I know that a square number is the result of the multiplication of a number by itself: consequently, 4, 9, etc., are square numbers, as they come from multiplying 2, 3, etc., by themselves.

Salv. Very well; and you also know that since the products are called squares, the factors are called sides or roots. On the other hand, the integers that don't consist of equal factors are not squares. Therefore, if I affirm that all integers, including both squares and non-squares, are more than squares alone, would I be telling the truth?

Simp. Sure.

Salv. However, I had to ask how many squares there are, the answer would be that there are as many squares as there are roots corresponding to them. Until each square will have its own root and every root will have its own square, every square will have no more than one root and every root cannot have more than one square.

Simp. It's true!

Salv. But if I ask how many roots there are, it cannot be denied that they are as many as the numbers, because every number is the root of a square and therefore the square numbers are as many as all the numbers because they are as many as their roots; and all the numbers are roots. At the beginning we said that there are many more numbers than squares, being most of them not squares. Not only: proportionally the number of squares decreases when we move to larger numbers. Up to a hundred, there are ten squares, i.e., the squares constitute 1/10 of all the numbers, while in ten thousand the fraction of squares is 1/100, and in one million only 1/1000. On the other hand, overall squares are as many as all the numbers.

Sagr. What is the conclusion of all this?

Salv. We can only conclude that the totality of numbers is infinite, that the number of squares is infinite, and also that the number of their roots is infinite. Neither is the number of squares less than the totality of all numbers, nor the latter greater than the former. Finally, the attributes "equal", "greater than" and "less than" do not apply to infinite quantities, but only to finite quantities. Therefore, when Simplicio introduces lines of different lengths and asks me how it is possible that the longer ones do not contain more points than the shorter, I answer him that a line does not contains more, less or just as many points than another one, but each line contains an infinite number. Can't I just do like I did for the roots and the squares and place more points in a line than in the other still maintaining an infinite number in both?

So much for the first difficulty.

Sagr. Please stop for a moment to let me add to what has been just said an idea that comes to my mind. If what we have said so far is correct, it seems to me impossible not only to affirm that an infinite number is greater than another infinite number, but also to say that it is greater than a finite number. Indeed if the infinite number were greater, for example, than a million, it would follow that, passing from the million to larger numbers, it would approach infinity. But this is not the case: rather, on the contrary, the greater the number, the greater is the distance from the infinite because the larger is the number and the fewer are the squares it contains, while in the infinite number the squares are not less than the totality of all the numbers, as we have just agreed.

Hence the approach to greater and greater numbers means a departure from infinity.

Salv. Thus from your ingenious argument we would conclude that the attributes "greater," "less," and "equal" have no place, neither in comparing infinite quantities with each other nor in comparing infinite with finite quantities. 80

I pass now to another consideration. Since lines and any extensions of them are divisible into parts that in turn are divisible to infinity, I am convinced that these lines are made up of infinite indivisible quantities. After all, a division that can be continuing without end presupposes that the parts are infinite, otherwise it would end; and if the parts are in infinite number, we can conclude that they aren't finite in size, because an infinite number of finite quantities would give an infinite magnitude. So a continuous quantity consists of an infinite number of indivisible parts.

Simp. But if we can carry on indefinitely the division into finite parts why should we need to introduce non-finite parts?

Salv. The fact that one is able to continue, without end, the division into finite parts, makes it necessary to regard the quantity as composed of an infinite number of immeasurably small elements. But going to the point I ask you: is a continuum made up of a finite or of an infinite number of finite parts?

Simp. My answer is that their number is both infinite and finite: potentially infinite but actually finite. Potentially infinite before division, and actually finite after division, because parts cannot exist in a body which is not yet divided or at least marked out; if this is not done, they exist potentially.

Salv. So we could not say that a twenty palms long line contains twenty lines of a palm each if not after the division into twenty equal parts: before the division, it contains them only potentially. Let's accept this point of view, and now please tell me: when the division is done, does the total length increase, diminish, or stay the same?

Simp. It neither increases nor diminishes.

Salv. That is my opinion too. Therefore the finite parts in a continuum, whether actually or potentially present, do not make the quantity either larger or smaller; but it is clear that, if the number of finite parts actually contained in the whole is infinite in number, they will make the magnitude infinite. Hence the number of finite 81 parts, although existing only potentially, cannot be infinite unless the magnitude containing them is infinite; conversely, if the magnitude is finite it cannot contain an infinite number of finite parts, neither actually nor potentially.

Sagr. How then is it possible to divide a continuum without limit into parts which are themselves capable of subdivision?

Salv. Your distinction between actual and potential makes it possible what appears to be impossible. But I shall try to reconcile these matters in another way; and to the question whether the parts of a limited continuum are finite or infinite in number I will, contrary to the opinion of Simplicio, answer that they are neither finite nor infinite.

Simp. This answer would never have occurred to me, since I didn't think it existed a middle ground between the finite and the infinite, such that the division between finite and infinite was incomplete or defective.

Salv. So it seems to me. If we talk about discrete quantities I think there is, between finite and infinite quantities, an intermediate term that corresponds to every assigned number, so that, to the question if the parts of a continuum are finite or infinite, the best answer is to say that they are neither finite nor infinite, but correspond to each assigned number. For this to be possible, it is necessary that they are not included in a limited number; nor can they be infinite, because no assigned number is infinite: and so, depending on whoever asks, we could divide any line into one hundred, one thousand, one hundred thousand finite parts, or any another number he likes, as long as it is not infinite. I therefore concede to philosophers that the continuum contains as many finite parts as they please and I concede also that it contains them, either actually or potentially, as they may like; but I must also add that as a line ten fathoms long contains ten lines of one fathom each and forty lines each of one cubit and eighty lines each of half a cubit, etc., so it contains an infinite number of points. You can call them actual or potential, as you like: on this detail, Simplicio, I stay with your opinion and with your judgment.

Simp. I admire what you say, but I'm afraid that this parallelism between the points and the finite parts contained in a line is not coherent, and that it will not be so easy to divide a given line into an infinite number of points, as it will not be for the philosophers to divide it into ten fathoms or forty cubits: not only that, but such a division is entirely impossible to achieve in practice, so this will remain one of those potential things that cannot actually be implemented.

Salv. The fact that something is feasible only with large effort, or spending much time, does not make it impossible. Nor would it be easy to divide a line into a thousand parts, and even less into 937 or any other large prime number. But if I could make this division as easy as it would be for another person to divide the line into forty parts, then would you concede the possibility of such a division?

Simp. I really like your method of dealing with the subject, often with great irony. I answer that it would seem to me more than sufficient, as long as dividing a line into points is not more laborious than dividing it into a thousand parts.

Salv. Now I want to tell you something that may surprise you about the possibility of dividing a line into an infinite number of points following the same procedure that is used in dividing it into forty, sixty or a hundred parts, that is dividing it subsequently into two, four, etc. Whoever thinks that, by following this method, one can reach an infinite number of points, is wrong, because if this process could be repeated to eternity and there would still be a finite number of parts.

Indeed with such a method we are far from reaching the goal of indivisibility; on the contrary, we move away from it, and while somebody believes that by continuing to divide and multiply the multitude of parts one approaches infinity, in my opinion, instead, we are moving further and further away. In the preceding discussion we concluded that, in an infinite number, it is necessary that the squares and cubes should be as numerous as the totality of the natural numbers, because both are as numerous as their roots which constitute the totality of the natural numbers. We have seen that the more numbers are taken, squares (1, 4, 9, 16, 25, 36, 49, ...) become less frequent, and cubes (1, 8, 27, 64, ..) even less frequent. Thus it is clear that the larger the numbers we consider, the farther we recede from the infinite; hence it follows that, since this process carries us farther and farther from our goal, if any number can be said to be infinite, it must be unity. Unity indeed satisfies all those conditions which are requisite for an infinite number; I mean that unity contains in itself as many squares as there are cubes and natural numbers.   *83*

Simp. I do not understand how to interpret this statement.

Salv. This statement leaves no doubt, because the number 1 is square (the square of one), cube (the cube of one), and fourth power (the fourth power of one). There is no property essential to all the squares or cubes that are not satisfied by unity. For example, a property satisfied by two squares is that there is an integer number which is mean proportional[g] between them. Take as a first term any square and as a second term the number one, you will always find a mean proportional. Between 9 and 1 the mean proportional is 3, while between 4 and 1 the mean proportional is 2, and between 9 and 4 the mean proportional is 6. A characteristic property of the cubes is instead that of having between them two mean proportionals. Take for example 8 and 27: between these two numbers there are 12 and 18, which are part of a geometric progression. Taking as extremes 1 and 8, we have 2 and 4, while taking 1 and 27 as extremes, the numbers in the sequence will be 3 and 9. We can therefore conclude that there is no other infinite number beside the unit. These are wonders that exceed the capacity of imagination, and must make us reflect on how much we are wrong to discuss the infinite with the same methods we use for finite quantities. The finite has no connection with the infinite, and therefore we need to introduce new methods to discuss the infinite itself.

In this regard, I want to tell a fact which now comes to mind, which explains the substantial difference in passing from a finite quantity to infinity.

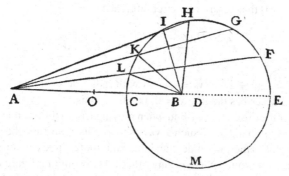

---

[g] The term "mean proportional" is equivalent to "geometric mean".

84-
85;90-
91

Let's take a segment AB of arbitrary length, and mark on it a point C that divides it into unequal parts. We draw all the pairs of segments with an extreme respectively in A and in B, whose lengths have the same proportion as AC and BC have, and which join in a point. The points where these segments join describe a circumference. For example, taking two segments AL and BL which have the same ratio with each other as AB with BC, and taking other pairs of segments AK, BK; AI, BI; AH, HB; AG, GB; AF, FB; AE, EB, then the points L, K, I, H, G, F, E all fall onto the same circumference. The circle described in this way will always be larger as C approaches the midpoint of the AB segment, which we call O, and will always be smaller as point C approaches B; with the law just described you can draw circles of arbitrary size: smaller than the pupil of an eye of a flea or larger of the equator of the celestial sphere.

Now, if we consider all the circles described passing through points included in the OB segment, we see that these become immense in the points close to O. Imagining to use precisely O with the same rule, namely that the segments obtained by joining it first with A and then with B keep the same proportion that AO maintains with OB, which line will be drawn? The circumference of a circle greater than all the others, and therefore infinite, is a straight line perpendicular to BA, drawn from the point O and conducted to infinity without ever reuniting its last extremity with the first. The circle larger than all, and therefore infinite, can no longer return to its initial point, and a straight line infinite line is the circumference of this infinite circle.

Indeed, calling $2L$ the length of the horizontal segment AB, $(0,0)$ the coordinates of the central point O, and the distance $|CO|$, we will have that the *geometric locus* of the points is described by the equation

$$\frac{x^2 + 2xL + L^2 + y^2}{x^2 - 2xL + L^2 + y^2} = \left(\frac{L+d}{L-d}\right)^2 \equiv 1 + \delta,$$

where δ is a positive quantity if $d > 0$. Thus

$$4xL - \delta(x^2 - 2xL + L^2 + y^2) = 0,$$

with $\delta \geq 0$. If $\delta \neq 0$ this describes a circumference

$$x^2 - \frac{2L(\delta + 2)}{\delta}x + y^2 + L^2 = 0$$

centered in $(x_c = L(\delta + 2)/\delta, \ y_c = 0)$ and with radius R such that $R^2 = x_c^2 - L^2$, while if $\delta = 0$ it describes the line $x = 0$, i.e., the y axis.

Now consider the difference we find when moving from a finite circle to an infinite one: this changes its character in such a way as to lose its nature and the possibility of existing as a circle. Rather, we understand that an infinite circle cannot exist; and for the same reason an infinite sphere or any other body or infinite figure cannot exist. Now, what about this transition from the finite to the infinite? And why should we feel greater revulsion than when we looked for the infinite and we found it in unity?

If we break a solid in many parts and reduce them to a very fine powder, until we arrive to its infinitely small indivisible atoms, why we could not say that this solid was reduced to a single continuum, perhaps fluid like water or mercury or even a liquefied metal? Don't we see that stones liquefy and become glass, and the glass itself under a strong heat becomes more fluid than water?

Sagr. Should we then believe that substances become fluid in virtue of being resolved into their infinitely small indivisible components?

Salv. I cannot find a better explanation for some phenomena, including the following one. When I take a hard body, be it stone or metal, and with a very hard tool I divide it into a very fine and impalpable powder, the finest particles, taken one by one, are so small to be imperceptible. But they have nonzero size anyway, and they can be counted. Accumulated together they support each other, and if we make an excavation, the cavity will remain and the particles around it will not rush to fill it. If shaken the particles come to rest immediately after the external disturbing agent is removed; the same effects are observed in all piles of larger and larger particles, of any shape, even if spherical, as is the case with piles of millet, wheat, lead shot, and every other material. But if we attempt to discover such properties in the water we do not find them; for when once heaped up it immediately flattens out unless held up by some vessel or other external retaining body; when hollowed out it quickly rushes in to fill the cavity; and when disturbed it fluctuates for a long time and sends out its waves through great distances. Seeing that water is less firm than the finest powder (in fact has no consistence whatever), we may, it seems to me, very reasonably conclude that the smallest particles into which it can be resolved are quite different from finite and divisible particles; indeed they might be indivisible. Gold and silver when pulverized with acids more finely than is possible with any mechanical means still remain powders, and do not become fluids until the finest particles of fire or of the rays of the Sun dissolve them, as I think, into their ultimate, indivisible, and infinitely small components.

Sagr. This phenomenon of the power of light which you mention is one which I have many times remarked with astonishment. I have, for instance, seen lead melted instantly by means of a concave mirror only three hands in diameter. Hence I think that if the mirror were very large, well polished and of a parabolic figure, it would just as readily and quickly melt any other metal, seeing that the small mirror, which was not well polished and had only a spherical shape, was able so energetically to melt lead and burn every combustible substance. Such effects make credible to me the marvels accomplished by the mirrors of Archimedes.

Salv. Speaking of the effects produced by the mirrors of Archimedes, it was his own books (which I had already read and studied with infinite astonishment) that rendered credible to me all the miracles described by various writers. And if any doubt had remained the book which Father Bonaventura Cavalieri has recently published on the subject of the burning glass and which I have read with admiration would have removed the last difficulties.

Sagr. When I saw this treatise I read it with wonder and pleasure and just this reading confirmed to me the idea, which I already had, about the fact that Cavalieri was destined to become one of the leading mathematicians of our time. But going back to talking about the wonderful effect of solar rays in liquefying metals, must we believe that this operation takes place without necessity of motion, or that is characterized by the most rapid of all motions?

Salv. We observe that combustion and dissolution are accompanied by very rapid motion. The actions of lightning and of the gunpowder used in mines are accompanied as well by fast movement. Bellows make combustion faster by increasing the speed of air. Hence I could not explain the action of light without admitting the intervention of the movement, and for truth the fastest of movements.

Sagr. How large is the speed of light? Its propagation may perhaps be instantaneous? Or does it take time, like other movements? Can we measure it experimentally?

Simp. Everyday experience shows that the propagation of light is instantaneous; for when we see a piece of artillery fired, at a great distance, the flash reaches our eyes without lapse of time; but the sound reaches the ear only after a noticeable interval.

Sagr. The only thing I am able to infer from this familiar bit of experience is that sound, in reaching our ear, travels more slowly than light; it does not inform me whether the coming of the light is instantaneous or whether, although extremely rapid, it still occupies time. An observation of this kind tells us nothing more than one in which it is claimed that "As soon as the Sun reaches the horizon its light reaches our eyes"; but who will assure me that these rays had not reached this limit earlier than they reached our vision?

Salv. The fact that these observations and other similar ones do not lead to conclusions made me think if it could be ascertained, without falling into error, if the propagation of light is really instantaneous. The comparison with the speed of sound shows that the propagation of light is at least very fast. I devised the following experiment.

Let each of two persons take a light contained in a lantern, such that by the interposition of the hand, the one can shut off or admit the light to the vision of the other. Next, let them stand opposite each other at a distance of a few cubits and practice until they acquire such skill in uncovering and occulting their lights that the moment one sees the light of his companion he will uncover his own. After a few trials the response will be so prompt that without sensible error the uncovering of one light is immediately followed by the uncovering of the other, so that as soon as one exposes his light he will instantly see that of the other. Having acquired skill at this short distance let the two experimenters, equipped as before, take up positions separated by a distance of two or three miles and let them perform the same experiment at night, noting carefully whether the exposures and occultations occur in the same time as at short distances; if they do, we may safely conclude that the propagation of light is instantaneous; but if time is required at a distance of three miles which, considering the going of one light and the coming of the other, amounts to six, then the delay ought to be easily observable. If the experiment is to be made at still greater distances, say eight or ten miles, telescopes may be employed, each observer adjusting one for himself at the place where he is to make the experiment at night; then although the lights are not large and are therefore invisible to the naked eye at so great a distance, they can readily be covered and uncovered since by aid of the telescopes, once adjusted and fixed, they will become easily visible.

Sagr. Very smart idea, indeed. Tell me what you found.

Salv. Unfortunately, I did the test only from a small distance, less than a mile, and for that I cannot discern if the appearance of the opposite light was really immediate or not. Anyhow if it is not immediate it is at least very fast, and I would compare it to the movement of lightning which we see in the clouds far tens of miles. We can see *89* the beginning of the light trail in a specific place in the clouds, but it immediately expands into the space that surrounds it. This seems to me an argument to conclude that propagation takes at least some time, since if the lighting were instantaneous we should not be able to distinguish its origin from the subsequent propagation.

But what trouble are we getting ourselves into little by little? Among all these speeches on the vacuum, on the indivisible, on the infinite and instantaneous movements, can we ever reach a safe haven?

Sagr. These matters lie far beyond our capability of understanding. Just think: when we seek the infinite among numbers we find it in unity; that which is ever divisible is derived from indivisibles; the vacuum is found inseparably connected with the plenum; indeed the views commonly held concerning the nature of these matters are so reversed that even the circumference of a circle turns out to be an infinite straight line.

Now we can try to satisfy Simplicio's desire by showing him how resolving a line in its infinite points is not only possible, but not even more difficult than resolving it in a finite number of parts. I hope, however, Simplicio, that you won't ask me to separate the points one from the other and show them one by one on this sheet of paper. In the same way, I am satisfied when a line is folded into a square or a hexagon *92* and I don't ask that the sides be actually separated.

Simp. Absolutely.

Salv. If the change which takes place when you bend a line at angles so as to form now a square, now an octagon, now a polygon of forty, a hundred or a thousand angles, is sufficient to bring into actuality the four, eight, forty, hundred, and thousand parts which, according to you, existed at first only potentially in the straight line, may I not say, with equal right, that, when I have bent the straight line into a polygon having an infinite number of sides (i.e., into a circle), I have reduced to actuality that infinite number of parts which you claimed were contained in it only potentially? Nor can one deny that the division into an infinite number of points is just as truly accomplished as the one into four parts when the square is formed or into a thousand parts when the "millagon" is formed; indeed in such a division the same conditions are satisfied as in the case of a polygon of a thousand or a hundred thousand sides. Such a polygon laid upon a straight line touches it with one of its sides, i.e., with one of its hundred thousand parts; while the circle which is a polygon of an infinite number of sides touches the same straight line with one of its sides which is a single point different from all its neighbors and therefore separate and distinct in no less degree than is one side of a polygon from the other sides. And just as a polygon, when rolled along a plane, marks out upon this plane, by the successive contacts of its sides, a straight line equal to its perimeter, so the circle rolled upon such a plane

93 also traces by its infinite succession of contacts a straight line equal in length to its own circumference. I don't know if the learned Peripatetics,[h] to which I grant as true the concept that the continuum is divisible into parts in turn always divisible in such a way that continuing with successive subdivisions an end would never be reached, grant me that, even if none of their divisions is the last, one last and what really solves the line in indivisible infinities really exists. But employing the method that I proposed, that of distinguishing and solving in one shot the infinite and all its parts, I believe they should be satisfied and should accept the composition of the continuous by indivisible atoms. Especially since this is probably the most direct way to extricate ourselves from many complicated problems. One of these, in addition to the already mentioned problem of the consistency of the parts of solids, is the understanding of rarefaction and condensation, avoiding being forced from the first to admit the existence of empty spaces, and from the second to the interpenetration of bodies. Both interpretations involve contradictions that seem to me to be cleverly avoided by assuming the composition by indivisibles.

Simp. I hardly know what the Peripatetics would say since the views advanced by you would strike them as mostly new, and as such we must consider them. It is however not unlikely that they would find answers and solutions to these problems which I, due to lack of time and critical capacity, am at present unable to solve. Leaving this aside for the moment, I should like to hear how the introduction of these indivisible quantities helps us to understand contraction and expansion avoiding at the same time the vacuum and the penetrability of bodies.

Sagr. I really want to listen to you, so this problem seems so obscure to me. I would like also to understand, as Simplicio reminded us, the arguments of Aristotle against the vacuum, and then the solution that you propose, since you admit what he denies.

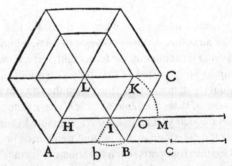

Salv. I'll do both. At first, with regard to rarefaction, let's think about the line described by the small inner circle we talked about earlier during a rotation of that larger—a line longer than the circumference of the small circle. To explain the contraction, we note that, during each rotation of the smaller circle, the larger one

---

[h] By extension the term indicates here the contemporaries of Galilei followers of Aristotle, who about two thousand years first held his lectures while walking in the avenue of the Athens high school, called Peripatos.

describes a straight line, that is shorter than its circumference. To better understand, let us consider what happens in the case of polygons.   *94*

Employing a figure similar to the earlier one, let's construct the two hexagons, ABC and HIK, about the common center L, and let them roll along the parallel lines HOM and ABc. Now holding the vertex I fixed, allow the smaller polygon to rotate until the side IK lies upon the parallel, during which motion the point K will describe the arc KM, and the side KI will coincide with IM. Let us see what, in the meantime, the side CB of the larger polygon has been doing. Since the rotation is about the point I, the terminal point B, of the line IB, moving backward, will describe the arc Bb underneath the parallel cA so that when the side KI coincides with the line MI, the side BC will coincide with bc, having advanced only through the distance Bc, but having retreated through a portion of the line BA which subtends the arc Bb. If we allow the rotation of the smaller polygon to go on, it will traverse and describe along its parallel a line equal to its perimeter; the larger one instead will traverse and describe a line shorter than its perimeter by as many times the length bB as there are its sides less one. This line is approximately equal to that described by the smaller polygon exceeding it only by the distance bB. Here now we see, without any difficulty, why the larger polygon, when carried by the smaller, does not measure off with its sides a line longer than that traversed by the smaller one; this is because a portion of each side is superposed upon its immediately preceding neighbor.

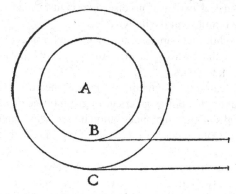

Let us next consider two circles, having a common center at A, and lying upon their respective parallels, the smaller being tangent to its parallel at point B; the larger, at the point C. When the small circle begins to roll, point B does not remain   *95* quiet for a while so as to allow BC to move backward and carry point C, as it did in the case of polygons, where the point I remained fixed until the KI side coincided with MI. In the case of circles, we must observe that the number of sides is infinite. In the case of the polygon, the vertices remain stationary during an interval of time that is to the period of a complete rotation in the same relationship with which a side is to perimeter. Similarly, in the case of circles, the delay of each of the infinite number of vertices is simply instantaneous (an instant is a fraction of a non-zero time interval such as a point of a segment, which contains an infinite number of points). The retrogression of the sides of the larger polygon is not equal to the length of one of its sides, but simply to the excess of such a side on one side of the smaller polygon, and the net advance is equal to this smaller side; but in the circle the point C, during the instantaneous arrest of B, withdraws of a quantity equal to its excess on the B side, having a net progress equal to B itself. In short, the infinite number of indivisible sides of the greater circle with their infinite number of indivisible, realized backprints during the infinite number of instantaneous delays of the infinite number of vertices of the smaller circle, together with the infinite number of progressions, equal to the infinite number of sides in the smaller circle, they add up in a line equal to the one described by the smaller circle, a line that contains an infinite number of infinitely small overlaps, thus determining a thickening or a contraction overlapping or interpenetration of non-null parts. This result cannot be obtained in the case of a line divided into finite parts like the perimeter of any polygon, which when it is arranged in a straight line cannot be shortened except through the overlap and the interpenetration of its sides. This contraction of an infinite number of infinitely small parts without the interpenetration or overlap of finite parts, added to the expansion of a number infinite of indivisible parts through the interposition of indivisible voids is, in my opinion, the maximum that we can say about the contraction and rarefaction of bodies, unless we abandon them the idea of impenetrability of matter and we introduce empty spaces of non-zero dimensions.

96

If you found something useful in my reasoning, please use it; otherwise, consider it as a useless chat, and seek explanations elsewhere. I just want to repeat that we are dealing here with the infinite and the indivisible.

Sagr. Your idea is subtle and impresses me as new and strange; but I am not able to say whether nature actually behaves like this. Anyway, until I find a more satisfactory explanation, I will stick to yours.

Perhaps Simplicio can tell us something I haven't heard yet, for example how to justify the explanation that philosophers have given of the most difficult subject. Everything I have read so far about condensation it is so dense, and all that concerns expansion is so subtle, that my poor brain can neither penetrate the first nor seize the second.

Simp. I am very confused; because according to this theory an ounce of gold might be rarefied and expanded until its size would exceed that of the Earth, while the Earth, in turn, might be condensed and reduced until it would become smaller than a walnut, something which I do not believe; nor do I believe that you believe it. The arguments and demonstrations which you have advanced are mathematical, abstract, and far removed from concrete matter; and I do not believe that when applied to the physical and natural world these laws will hold.

97-98

Salv. I doubt you want me to show you the invisible, nor am I able to do it; but for what concerns what can be perceived by our senses, since you mentioned gold, can you comment on the immense possibility of expanding its parts?

I don't know if you've thought about how artisans proceed to prepare the golden threads, which are made of gold only on the surface, while inside they are made of silver. They take a silver cylinder or a rod half a cubit long and three or four fingers thick; they cover it with eight or ten beaten gold leaves, which as you know are so thin that they float in the air. Then they begin to pull with great force, pressing the cylinder through holes always smaller. After many steps they reduce it to the fineness of a woman's hair or even finer; however, it remains golden on the surface. I let you imagine the subtlety and expansion to which gold is subjected.

Simp. I don't see how this operation causes a thinning of the gold to justify your wonder. First, the ten leaves that make up the gilding represent a noticeable thickness; secondly, if silver grows in length, it equally decreases in thickness; and so one dimension compensates the other, and the surface does not increase to such point that to cover gold with silver it is necessary to give it a greater thickness than that of the former leaves.

Salv. You're wrong, Simplicio, because the surface with the same volume grows like the root square of length, and I can prove it to you.

Sagr. Please prove it if you think it's not too difficult.

Salv. Given a cylinder of radius r and height h (the height of the cylinder represents the length of the cylinder), the area S of the outer surface (i.e., without the bases) is

$$S = 2\pi r\, h \tag{1}$$

while the volume V is

$$V = \pi r^2 h = \frac{1}{4\pi}\frac{S^2}{h}. \tag{2}$$

Thus

$$S = \sqrt{4\pi h V}\,, \tag{3}$$

as we wanted to prove.

If we apply the results obtained to the case in question and suppose that the silver cylinder before stretching had a half-cubit length and a thickness of three or four inches, we will find that, when the thread has been reduced to the fineness of a hair and has been pulled to twenty thousand cubits length (and maybe more), the surface is increased two hundred times. As a result the ten gold leaves have now been laid out on a surface two hundred times larger, and the thickness of the gold which eventually covers the surface of so many cubits of a thread is not greater than one twentieth of that of an ordinary gold leaf. Now consider the degree of finesse you need have and if it can be conceived that this will happen in any other way than with an enormous expansion of the parts; also considers whether this experiment suggests that matter is composed of infinite small indivisible particles, a vision that is also proved by other more impressive and decisive examples.

Sagr. The demonstration was very simple and I liked it.   99

Salv. Since you are so passionate about these demonstrations, and they bring with them discrete advantages, I propose to you a theorem that answers an extremely interesting question. We have seen first what relations exist between cylinders with the same volume but of different height or length; let's see now what happens when the cylinders are equal in area but different in height, meaning for the area S that of the lateral surface, i.e., excluding the upper and lower bases.

The theorem, evident from equation (2), is: the volumes of cylinders having an
100 equal area of the lateral surfaces are inversely proportional to their heights.

This explains a phenomenon to which ordinary people always look with wonder.
With a piece of cloth we can make a flour sack, using a wooden base as usual. If the
fabric has one side longer than the other, the bag will be larger when the short side of
the fabric is used for the height and the long side is wrapped around the wooden base
rather than the other layout. So, for example, from a piece of fabric of six cubits along
one side and twelve long the other, a sack can be made that will hold more when the
side of twelve cubits is wrapped around at the wooden base, leaving the height of
the bag at six cubits, which when the side of six cubits comes wrapped at the base
making the bag twelve cubits high. According to what was first demonstrated, we
learn not only the general fact that a lot contains more than the other, but we get also
specific and particular information on how much more: for example as in proportion
to what the height of the bag decreases the content of this increase, and vice versa.
If we use the data provided (a fabric twice as long as the width), using the long side
for the height, the volume of the sack will be the half compared to what you would
have in the opposite arrangement.

101 *Sagr.* It is with great pleasure that we continue thus to acquire new and useful
information. But as regards the subject just discussed, I really believe that, among
those who are not already familiar with geometry, you would scarcely find four
persons in a hundred who would not, at first sight, make the mistake of believing that
bodies having equal surfaces would be equal in other respects. Speaking of areas,
the same error is made when one tries, as often happens, to determine the sizes of
various cities by measuring their boundary lines, forgetting that the circuit of one
may be equal to the circuit of another while the area of the one is much greater than
that of the other. And this is true not only in the case of irregular, but also of regular
surfaces, where the polygon having the greater number of sides always contains a
larger area than the one with the fewer sides, so that finally the circle which is a
polygon of an infinite number of sides, contains the largest area of all polygons of
equal perimeter. I remember with particular pleasure having seen this demonstration
when I was studying *The sphere* by Sacrobosco[i] with the help of a particularly learned
commentator.

102- *Salv.* Very true! I too came across the same passage which suggested to me a
103 method of showing how, by a single short demonstration, one can prove that the
circle has the largest area among all regular isoperimetric figures; and that, of other
figures, the one which has the larger number of sides contains a greater area than that
which has the smaller number.

Sagr. I am very fond of non-trivial propositions: please let us listen to your demonstration.

Salv. I can do it in a few words demonstrating the following

---

[i] It is a treatise on astronomy written in the thirteenth century by the English astronomer John
Holywood (Sacrobosco) who taught at the University of Paris. Galilei himself adopted it in Padua
as an elementary text on astronomy, probably when Sagredo was attending his lectures.

Theorem The area of a circle is mean proportional between two similar regular polygons of which one circumscribes and the other has the same perimeter of the circumference. Furthermore, the area of the circle is larger than that of any polygon that has the same perimeter. Finally, of these circumscribed polygons, that which has the larger number of sides has an area smaller than that which has the smaller number of sides; but, on the other hand, the polygon that has the greatest number of sides is that of the largest area among those that have the same perimeter.

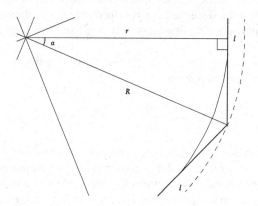

Referring to the figure,[9] let 2α be the angle at the center of each of the triangles that make up a regular polygon of n sides, each of length ℓ. The perimeter p_n is

$$p_n = n\ell = 2nR \sin \alpha = 2nr \frac{\sin \alpha}{\cos \alpha}. \qquad (4)$$

where r is the radius of the inscribed circle, and R is the radius of the circumscribed circumference. For calculating the area of a regular polygon is enough to multiply by n the area of the isosceles triangles that make it up. Therefore, since these triangles are based on one side and height on the radius of the inscribed circumference, the regular polygon of n sides has an area

$$A_n = n\frac{r\ell}{2} = nr^2 \frac{\sin \alpha}{\cos \alpha} = \frac{p_n^2}{4n} \frac{\cos \alpha}{\sin \alpha}. \qquad (5)$$

We want to show that, given a circle of radius r,

$$A^2 = (\pi r^2)^2 = A_e A_i,$$

where A is the surface of the circle, A_e is the surface of the outer (circumscribed) polygon, and A_i the area of the polygon with the same number of sides and with the same perimeter as the circumference. From (4) and (5) we have

$$A_i = \frac{p_n^2}{4n}\frac{\cos\alpha}{\sin\alpha} = \frac{(2\pi r)^2}{4n}\frac{\cos\alpha}{\sin\alpha} = \frac{\pi^2 r^2}{n}\frac{\cos\alpha}{\sin\alpha}$$

$$A_e = nr^2\frac{\sin\alpha}{\cos\alpha},$$

and thus $A_e A_i = \pi^2 r^4$, q.e.d.

The second statement is that the area of the circle is greater than that of any polygon which has the same perimeter, descends directly from this: since $A_e > A$ for each number of sides since the circumscribed polygon contains the circle, then, being A the geometrical mean between A_e e A_i, we have $A > A_i$ irrespective of the number n of sides.

As for the last two statements, from (5) and considering that $\alpha = \pi/n$, we have, for a generic circumscribed polygon,

$$A_n = nr^2\frac{\sin\alpha}{\cos\alpha} = \pi r^2\frac{\tan\alpha}{\alpha}.$$

Always $\alpha < \tan\alpha$, but the arc gets closer and closer to the chord as the number of sides of the polygon increases, and therefore the ratio approximates unity: we have shown that the polygon that has the greatest number of sides is that of the smallest area among those circumscribed. Finally, always from Eq. (5), we obtain that. fixing the perimeter p_n of a polygon,

$$A_n = \frac{p_n^2}{4\pi}\frac{\alpha}{\tan\alpha};$$

similarly to when done before we have shown that the polygon that has the greatest number of sides is that of the largest area among those that have the same perimeter.

Sagr. A very smart and elegant demonstration! But how did we move to geometry while we were discussing the important objections raised by Simplicio, especially that which refers to the condensation that seems to me particularly difficult to answer?

Salv. If contraction and expansion consist of contrary motions, one ought to find for each great expansion a correspondingly large contraction. But our surprise is increased when, every day, we see enormous expansions taking place almost instantaneously. Think about what a tremendous expansion occurs when a small quantity of gunpowder flares up into a vast volume of fire! Think also of the almost limitless expansion of the light which it produces! Imagine the contraction which would take place if this fire and this light were to reunite, which, indeed, is not impossible since only a little while ago they were located together in this small space. You will find, upon observation, a thousand such expansions, since they are more frequent than contractions. We can take wood and see it go up in fire and light, but we don't see them recombine to form wood; we see fruits and flowers and a thousand other solid bodies dissolve largely into odors, but we do not observe these fragrant atoms coming together to form fragrant solids.

But where the senses fail us reason must step in; it will enable us to understand the motion involved in the condensation of extremely rarefied and tenuous substances just as clearly as that involved in the expansion and dissolution of solids. Moreover, we are trying to find out how it is possible to produce expansion and contraction in bodies that are capable of such changes without introducing vacua and without giving up the impenetrability of matter; but this does not exclude the possibility of there being materials that possess no such properties and do not, therefore, carry with them consequences which you call inconvenient and impossible. And finally, Simplicio, I have, for the sake of you philosophers, tried to find an explanation of how expansion and contraction can take place without our admitting the penetrability of matter and introducing vacua, properties which you deny and dislike; if you were to admit them, I should not oppose you so vigorously. Now either admit these difficulties or accept my views or suggest something better.

Sagr. I quite agree with the peripatetic philosophers in denying the penetrability of matter. As to the vacua I would like, please, to hear from Simplicio a thorough discussion of Aristotle's demonstration in which he opposes them, and what you, Salviati, have to say in reply.

Simp. So far as I remember, Aristotle opposes the ancient view that vacuum is a necessary prerequisite for motion and that the latter could not occur without the former. In opposition to this view, Aristotle shows that it is precisely the phenomenon of motion, as we shall see, which renders untenable the idea of the vacuum.[10] His method is to divide the argument into two parts. He first supposes bodies of different weights to move in the same medium; then supposes, one and the same body to move in different media. In the first case, he supposes bodies of different weight to move in one and the same medium with different speeds which stand to one another in the same ratio as the weights; so that, for example, a body which is ten times as heavy as another will move ten times as rapidly as the other. In the second case he assumes that the speeds of one and the same body moving in different media are in inverse ratio to the densities of these media; thus, for instance, if the density of water were ten times that of air, the speed in air would be ten times greater than in water. From this second supposition, he shows that, since the tenuity of vacuum differs infinitely from that of any medium filled with matter however rare, any body which moves in a plenum through a certain space in a certain time ought to move through vacuum instantaneously; but the instantaneous motion is impossible, therefore the vacuum cannot exist.

Salv. We see that the argument is specifically against those who believed that vacuum is needed for motion. Now if I admit the argument to be conclusive and concede also that motion cannot take place in vacuum, the assumption of a vacuum considered not with reference to motion, is not invalidated. But to tell you what the ancients might possibly have replied and to better understand Aristotle's argument, I think we may contradict both hypotheses. Regarding the first, I doubt that Aristotle ever tried to see if two stones of different weights (one, for example, ten times the other) dropped at the same time by one same height (for example of a hundred cubits) acquired a speed so different that upon arrival of the largest on the ground the other was still far away.

106

Simp. Yet he shows that he experienced it from his words, because he says "We see that the heaviest..."[11] that verb "to see" indicates that he has made the experiment.

107   *Sagr.* But I, Simplicio, have made the test, and can assure you that a cannonball weighing one or two hundred pounds, or even more, will not reach the ground by as much as a span ahead of a musket ball weighing only half a pound, provided both are dropped from a height of 200 cubits.

Salv. But, even without further experiment, it is possible to prove clearly, by means of a short and conclusive argument, that a heavier body does not move more rapidly than a lighter one provided both bodies are of the same material and in short such as those mentioned by Aristotle. But tell me, Simplicio, whether you admit that each falling body acquires a definite speed fixed by nature, a velocity which cannot be increased or diminished except by the use of force or resistance.

Simp. There is no doubt but that the same body moving in a single medium has a fixed velocity which is determined by nature and which cannot be increased except by the addition of impetus or diminished except by some resistance that retards it.

Salv. If then we take two bodies whose natural speeds are different, it is clear that on uniting the two, the more rapid one will be partly retarded by the slower, and the slower will be fastened by the swifter. Don't you agree with me in this opinion?

Simp. I agree.

Salv. But if this is true, and if a large stone moves with a speed of, say, eight grades, while a smaller one moves with a speed of four, then when they are united, the system will move with a speed less than eight; but the two stones when tied together make a stone larger than that which before moved with a speed of eight. Hence the heavier body moves with less speed than the lighter; an effect which is *108*   contrary to your supposition. Thus, you see, it can't true that the heavier mobile moves faster than the lightest.

Simp. I feel in trouble, because I believe that the smaller stone, if combined with the larger one, adds weight to the latter, and adding weight, it should also add speed, or at least not take it off.

Salv. Here again you are in error, Simplicio, because it is not true that the smaller stone adds weight to the larger.

Simp. This is, indeed, quite beyond my comprehension.

Salv. It is necessary to distinguish between heavy bodies in motion and the same bodies at rest. A large stone placed on a scale not only acquires additional weight by having another stone placed upon it, but even by the addition of a handful of hemp its weight is augmented six to ten ounces according to the quantity of hemp. But if you tie the hemp to the stone and allow them to fall freely from some height, do you believe that the hemp will press down upon the stone and thus accelerate its motion, or do you think the motion will be retarded by a partial upward pressure? One always feels the pressure upon his shoulders when he prevents the motion of a load resting upon him; but if one descends just as rapidly as the load would fall how can it gravitate or press upon him? Don't you see that this would be the same as trying to strike a man with a spear when he is running away from you with a speed which is equal to, or even greater, than that with which you are striking him? You must therefore conclude that, during free and natural fall, the small stone does not press upon the larger and consequently does not increase its weight as it does when at rest.

Simp. But what if we place the larger stone upon the smaller?

Salv. Its weight would be increased if the larger stone moved more rapidly; but *109* we have already concluded that when the small stone moves more slowly it retards to some extent the speed of the larger, so that the combination of the two, which is a heavier body than the larger of the two stones, would move less rapidly, a conclusion which is contrary to your hypothesis. We infer therefore that large and small bodies move with the same speed provided they have the same specific weight.

Simp. Your reasoning seems convincing to me; however it seems to me difficult to believe that a lead drop can move as fast as a cannonball.

Salv. Why not say a grain of sand as rapidly as a grindstone? But, Simplicio, I trust you will not follow the example of many others who divert the discussion from its main intent and criticize some statement of mine which lacks a hair's breadth of the truth and, under this hair, hide the fault of another which is as big as a ship's cable. Aristotle says that an iron ball of one hundred pounds falling from a height of one hundred cubits reaches the ground before a one-pound ball has fallen a single cubit; I say that they arrive at the same time. You find, on making the experiment, that when the heavier has reached the ground, the other is short of it by two inches; now you would not hide behind these two fingers the ninety-nine cubits of Aristotle, nor would you mention my small error and at the same time pass over in silence his very large one.

Aristotle declares that bodies of different weights, in the same medium, travel (in so far as their motion depends upon gravity) with speeds proportional to their weights. He illustrates this by using bodies in which it is possible to perceive the pure and unadulterated effect of gravity, eliminating other considerations, for example influences that are greatly dependent upon the medium which modifies the effect of gravity alone. Thus we observe that gold, the densest of all substances, when shaped out into a very thin leaf, floats through the air; the same thing happens with stone when ground into a very fine powder. But if you wish to maintain the general proposition you will have to show that the same ratio of speeds is preserved in the case of all heavy bodies, and that a stone of twenty pounds moves ten times as rapidly *110* as one of two; I claim instead that this is false and that, if they fall from a height of fifty or a hundred cubits, they will reach the ground at the same moment.

Simp. Perhaps the result would be different if the fall took place not from a few cubits but from some thousands of cubits.

Salv. If this were what Aristotle meant you would burden him with another error which would amount to a falsehood; because, since there is no such sheer height available on Earth, it is clear that Aristotle could not have made the experiment. Yet he wishes to give us the impression of his having performed it when he speaks of such an effect as one which we see.

Simp. In fact, Aristotle does not employ this principle, but uses the other one which is not, I believe, subject to these same difficulties.

Salv. But the one is as false as the other; and I am surprised that you yourself don't see the fallacy and that you do not perceive that if it were true that, in media of different densities and different resistances, such as water and air, one and the same body moved in the air more rapidly than in water, in proportion as the density of water is greater than that of air, then it would follow that any body which falls through air ought also falls through water. But this conclusion is false: many bodies which descend in air not only do not descend in water, but actually rise.

Simp. Aristotle discusses only those bodies which fall in both media, not those which fall in the air but rise in water.

Salv. The arguments which you advance for Aristotle are such as he himself would have certainly avoided. But tell me if between the density of water, or what delays the motion more, and that of air, which delays it less, there is a ratio, and tell me how much it is.

Simp. Sure it exists, and we suppose that it is a factor of 10: the speed of a body descending in water will be ten times smaller than in the air.

Salv. I shall now take one of those bodies which fall in air but not in water, say a wooden ball, and I shall ask you to assign to it any speed you please for its descent through air.

Simp. Let us suppose it moves with a speed of twenty.

Salv. Very well. Then it is clear that this speed bears to some smaller speed the same ratio as the density of water bears to that of air; and the value of this smaller speed is two. So that if we follow exactly the assumption of Aristotle we ought to infer that the wooden ball which falls in air, a substance ten times less resisting than water, with a speed of twenty grades would fall in water with a speed of two, instead of coming to the surface from the bottom as it does; unless perhaps you wish to reply, which I do not believe you will, that the rising of the wood through the water is the same as its falling with a speed of two. But since the wooden ball does not go to the bottom, I think you will agree with me that we can find a ball of another material, not wood, which does fall in water with a speed of two.

Simp. No doubt we can; but it must be of a substance considerably heavier than wood.

Salv. That is it exactly. But if this second ball falls in water with a speed of two, what will be its speed of descent in the air? If you hold to the rule of Aristotle you must reply that it will move at the rate of twenty; but twenty is the speed which you yourself have already assigned to the wooden ball; hence this and the other heavier ball will each move through the air with the same speed. But now how does Aristotle harmonize this result with his other, namely, that bodies of different weight move through the same medium with different speeds—speeds which are proportional to their weights? How have these common and obvious properties escaped your notice?

Have you not observed that two bodies which fall in water, one with a speed a hundred times as great as that of the other, will fall in air with speeds so nearly equal that one will not surpass the other by as much as one hundredth part? Thus, for example, an egg made of marble will descend in water one hundred times more rapidly than a hen's egg, while in air falling from a height of twenty cubits the one will fall short of the other by less than four inches. In short, a heavy body sinking through ten cubits of water in three hours will traverse ten cubits of air in one or two pulse-beats; and if the heavy body is a ball of lead it will easily traverse the ten cubits of water in less than double the time required for ten cubits of air. And here, I am sure, Simplicio, you find no ground for difference or objection. We conclude, therefore, that the argument does not bear against the existence of vacuum; but if it did, it would only do away with vacua of considerable size which neither I nor, in my opinion, the ancients ever believed to exist in nature, although they might possibly be produced by force as may be gathered from various experiments whose description would here occupy too much time.

Sagr. Seeing that Simplicio is silent, I will say something. You have clearly demonstrated that bodies of different weights do not move in one and the same medium with velocities proportional to their weights, but that they all move with the same speed, understanding of course that they are of the same substance or at least of the same specific weight. This is certainly not true if they have different specific weights, for I hardly think you would have us believe a ball of cork moves with the same speed   *113* as one of lead; and again since you have clearly demonstrated that one and the same body moving through differently resisting media does not acquire speeds which are inversely proportional to the resistances, I am curious to learn what are the ratios actually observed in these cases.

Salv. These are interesting questions and I have thought much concerning them. Having established the falsity of the proposition that one and the same body moving through differently resisting media acquires speeds which are inversely proportional to the resistances of these media, and having also disproved the statement that in the same medium bodies of different weight acquire velocities proportional to their weights (understanding that this applies also to bodies which differ merely in specific weight), I then began to combine these two facts and to consider what would happen if bodies of different specific weight were placed in media of different resistances; and I found that the differences in speed were greater in those media which were more resistant, that is, less yielding. This difference was such that two bodies which differed scarcely at all in their speed through air would, in water, fall the one with a speed ten times as great as that of the other. Further, there are bodies which will fall rapidly in air, whereas if placed in water not only will not sink but will remain at rest or will even rise to the top: for it is possible to find some kinds of wood, such as knots and roots, which remain at rest in water but fall rapidly in air.

Sagr. I have often tried with patience to add grains of sand to a ball of wax until it should acquire the same specific weight as water and would therefore remain at rest in this medium. But with all my care I was never able to accomplish this. Indeed, I do not know whether there is any solid substance whose specific weight is, by nature, so nearly equal to that of water that if placed anywhere in water it will remain at rest.

Salv. In this, as in a thousand other operations, humans are surpassed by animals. In this problem of yours, one may learn much from fish, which are very skillful in maintaining their equilibrium n in waters with different densities. So perfectly indeed   *114* can fish keep their equilibrium that they are able to remain motionless in any position. This they accomplish, I believe, by means of an apparatus especially provided by nature, namely, a bladder located in the body and communicating with the mouth by means of a narrow tube through which they are able, at will, to expel part of the air contained in the bladder. By rising to the surface they can take in more air; thus they make themselves heavier or lighter than water at will and maintain equilibrium.

Sagr. By means of another device I was able to impress some friends to whom I had boasted that I could make up a ball of wax that would be in equilibrium in water. At the bottom of a vessel I placed some salt water and upon this some fresh water; then I showed them that the ball stopped in the middle of the water, and that, when pushed to the bottom or lifted to the top, would return to the middle.

Salv. This experiment is useful. For when physicians are testing the qualities of various waters, especially their specific weights, they employ a ball of this kind calibrated in such a way that, in certain water, it will neither rise nor fall. Then in testing another water, differing ever so slightly in density, the ball will sink if this water is lighter and rise if it is heavier. And so exact is this experiment that the addition of two grains of salt to six pounds of water is sufficient to make the ball rise to the surface from the bottom to which it had fallen. To illustrate the precision of this experiment and also to clearly demonstrate the lack of resistance of water to division, I wish to add that this notable difference in specific weight can be produced not only by solution of some heavier substance, but also by merely heating or cooling; and so sensitive is water to this process that by simply adding four drops of another water which is slightly warmer or cooler than the six pounds one can cause the ball to sink or rise; it will sink when the warm water is poured in and will rise upon the addition of cold water.[j] Now you can see how mistaken are those philosophers who ascribe to water viscosity or some other coherence of parts which offers resistance to separation and penetration.

Sagr. Concerning this question I have found many convincing arguments in a treatise[12] by our Academician; but there is one great difficulty of which I have not been able to get rid, namely, if there be no tenacity or coherence between the particles of water, how is it possible for those large drops of water to stand out in relief upon cabbage leaves without scattering or spreading out?

Salv. To begin with, let me confess that I do not understand how these large globules of water stand out and hold themselves up, although I know for sure, that it is not owing to any internal tenacity acting between the particles of water; whence it must follow that the cause of this effect is external. Besides the experiments already shown to prove that the cause is not internal, I can offer another which is very convincing. If the particles of water that sustain themselves in a heap, while surrounded by air, did so in virtue of an internal cause then they would sustain themselves much more easily when surrounded by a medium in which they exhibit less tendency to fall than they do in the air; such a medium would be any fluid heavier than air, as, for instance, wine: and therefore if some wine is poured about such a drop of water, the

---

[j] This principle is the basis for the operation of the so-called Galilean thermometer. The instrument consists of a glass cylinder containing a liquid and some glass cruets containing in turn some liquid; the ampoules have average densities different from each other and labels are hung on them indicating a temperature obtained from an appropriate calibration. When the device is in thermal equilibrium with the environment, a measurement of the temperature (if the temperature is within the *range* of instrument operation) is given by the number shown on the lowest of the remaining cruets floating.

wine might rise until the drop was entirely covered, without the particles of water, held together by this internal coherence, ever parting company. But this is not the case: as soon as the wine touches the water, the latter without waiting to be covered scatters and spreads out underneath the wine, as you can easily see if you use red wine. The cause of this effect is therefore external and is possibly to be found in the surrounding air. Indeed there appears to be a considerable antagonism between air and water as I have observed in the following experiment.   *116*

Having taken a glass globe that had a mouth of about the same diameter as a straw, I filled it with water and turned it mouth downwards. The water, although quite heavy and prone to descend, and the air, which is very light and disposed to rise through the water, refused, the one to descend and the other to ascend through the opening, but both remained stubborn and defiant. On the other hand, as soon as I apply to this opening a glass of red wine, which is almost inappreciably lighter than water, red streaks are immediately observed to ascend slowly through the water while the water with equal slowness descends through the wine without mixing, until finally the globe is completely filled with wine and the water has all gone down into the vessel below.

Simp. I laugh at the great antipathy which Salviati exhibits against the use of the word antipathy[k]; and can easily explain this difficulty.

Salv. Let this word antipathy be the solution to our difficulty if Simplicio likes so. Returning from this digression, let us again take up our problem. We have already seen that the difference of speed between bodies of different specific weights is most marked in those media which are the most resistant: thus, in a medium of mercury, gold does not merely sink to the bottom more rapidly than lead, but it is the only substance that will descend: all other metals and stones rise to the surface and float. On the other hand, the variation of speed in air between balls of gold, lead, copper, and other heavy materials is so slight that in a fall of 100 cubits a ball of gold would surely not outstrip one of copper by as much as four fingers. Having observed this I came to the conclusion that in a medium totally devoid of resistance all bodies would fall with at the same speed.

Simp. This is a remarkable statement, Salviati. But I shall never believe that even in vacuum, if motion in such a place were possible, a lock of wool and a small ball of lead can fall with the same velocity.

Salv. A little slower, Simplicio. I have already considered this matter and found the proper solution. Our problem is to find out what happens to bodies of different weight   *117*
moving in a medium devoid of resistance, so that the only difference in speed is that which arises from an inequality of weight. Since no medium except one entirely free from air and other bodies, be it ever so tenuous and yielding, can provide our senses the evidence we are looking for, and since such a medium is not available, we shall observe what happens in the rarest and least resistant media as compared with what happens in denser and more resistant media. Because if we find as a fact

---

[k] The concept of sympathy and its opposite of antipathy were used by peripatetics also in the context of physics, as a tendency to approach or to move away. For Hermetic philosophers (2nd century AD), sympathy and antipathy also influenced the relationship between humans and nature.

that the variation of speed among bodies of different specific weights is less and less when the medium becomes more and more yielding, and if finally in a medium of extreme tenuity, though not a perfect vacuum, we find that, in spite of great diversity of specific weight, the difference in speed is very small and almost inappreciable, then we are justified in believing that in the vacuum all bodies would fall with the same speed. Let us, in view of this, consider what takes place in the air, where for the sake of a definite figure and light material imagine an inflated bladder. The air in this bladder when surrounded by air will weigh little or nothing, since it can be only slightly compressed; its weight then is small being merely that of the skin, which does not amount to the thousandth part of a mass of lead having the same size as the inflated bladder. Now, Simplicio, if we allow these two bodies to fall from a height of four or six cubits, by what distance do you imagine the lead will anticipate the bladder? You may be sure that the lead will not travel three times, or even twice, as swiftly as the bladder, although you would have made it move a thousand times as rapidly.

Simp. It may be as you say during the first four or six cubits of the fall; but after the motion has continued a long while, I believe that the lead will have left the bladder behind not only six out of twelve parts of the distance but even eight or ten.

Salv. I quite agree with you and don't doubt that, in very long distances, the lead might cover one hundred miles while the bladder was traversing one; but, my dear Simplicio, this phenomenon which you adduce against my proposition is precisely the one which confirms it.

Let me once more explain that the variation of speed observed in bodies of different specific weights is not caused by the difference of specific weight but depends upon external circumstances and, in particular, upon the resistance of the medium, so that if this is removed all bodies would fall with the same velocity; and this result I deduce mainly from the fact which you have just admitted and which is very true, namely, that, in the case of bodies which differ widely in weight, their velocities differ more and more as the spaces traversed increase, something which would not occur if the effect depended upon differences of specific weight. Since these specific weights remain constant, the ratio between the distances traversed ought to remain constant whereas the fact is that this ratio keeps on increasing as the motion continues. Thus a very heavy body in a fall of one cubit will not anticipate a very light one by so much as the tenth part of this space; but in a fall of twelve cubits the heavy body would outstrip the other by one third, and in a fall of one hundred cubits by 90/100, etc.

Simp. Very well: but, following your line of argument, if differences of weight in bodies of different specific weights cannot produce a change in the ratio of their speeds, on the ground that their specific weights do not change, how is it possible for the medium, which also we suppose to remain constant, to bring about any change in the ratio of these velocities?

Salv. This objection is clever; and I must meet it. I begin by saying that a heavy body has an inherent tendency to move with a constantly and uniformly accelerated motion toward the common center of gravity, that is, toward the center of our Earth, so that during equal intervals of time it receives equal increments of momentum and velocity. This, you must understand, holds whenever all external and accidental hindrances have been removed; but of these, there is one which we can never remove, namely, the medium which must be penetrated and thrust aside by the falling body, i.e., air. This quiet, yielding, fluid medium opposes motion through it with a resistance *119* which is proportional to the rapidity with which the medium must give way to the passage of the body; which body, as I have said, is by nature continuously accelerated so that it meets with more and more resistance in the medium and hence it experiences a diminution in its rate of gain of speed until finally the speed reaches such a point and the resistance of the medium becomes so great that, balancing the weight, prevents any further acceleration and reduces the motion of the body to a uniform one which will thereafter maintain a constant speed. There is, therefore, an increase in the resistance of the medium, not on account of any change in its essential properties, but on account of the change in rapidity with which it must yield and give way laterally to the passage of the falling body which is being constantly accelerated.

Now seeing how great is the resistance which the air offers to the slight momentum of the bladder and how small that which it offers to the large weight of the lead, I am convinced that, if the medium were entirely removed, the advantage received by the bladder would be so great and that coming to the lead so small that their speeds would be equalized. Assuming this principle, that all falling bodies acquire equal speeds in a medium which, on account of vacuum or something else, offers no resistance to the speed of the motion, we shall be able accordingly to determine the ratios of the speeds of both similar and dissimilar bodies moving either through one and the same medium or through different space-filling, and therefore resistant, media. We may obtain this result by observing how much the weight of the medium detracts from the weight of the moving body, which weight is the means employed by the falling body to open a path for itself and to push aside the parts of the medium, something which does not happen in vacuum where, therefore, no difference in speed is to be expected from a difference of specific weight. And since it is known that the effect of the medium is to diminish the weight of the body by the weight of the medium displaced, we may accomplish our purpose by diminishing in just this proportion the speeds of the falling bodies, which in a non-resisting medium we have assumed to be equal. Thus, for example, imagine lead to be ten thousand times as heavy as air while ebony is only one thousand times as heavy. Here we have two substances *120* whose speeds of fall in a medium devoid of resistance are equal: but, when air is the medium, it will subtract from the speed of the lead one part in ten thousand, and from the speed of the ebony one part in one thousand, i.e., ten parts in ten thousand. While therefore lead and ebony would fall from any given height in the same interval of time, provided the retarding effect of the air was removed, the lead will, in air, lose in speed one part in ten thousand; and the ebony, ten parts in ten thousand. In other words, if the elevation from which the bodies start is divided into ten thousand parts, the lead will reach the ground leaving the ebony behind by as much as ten, or at least nine, of these parts. Is it not clear then that a leaden ball allowed to fall from a tower two hundred cubits high will outstrip an ebony ball by less than four inches? Now ebony weighs a thousand times as much as air but this inflated bladder only four times as much; therefore air diminishes the inherent and natural speed of ebony by one part in a thousand; while that of the bladder which, if free from hindrance, would be the same, experiences a diminution in air amounting to one part in four. So that when the ebony ball, falling from the tower, has reached ground, the bladder will have traversed only three quarters of this distance. Lead is twelve times as heavy as water; but ivory is only twice as heavy. The speeds of these two substances which, when entirely unhindered, are equal will be diminished in water, that of lead by one part in twelve, that of ivory by half. Accordingly, when the lead has fallen through eleven cubits of water the ivory will have fallen through only six. Employing this principle we shall, I believe, find a much closer agreement of experiment with our computation than with that of Aristotle.

In a similar manner we may find the ratio of the speeds of one and the same body in different fluids, not by comparing the different resistances of the media, but by considering the excess of the specific weight of the body above those of the media. Thus, for example, tin is one thousand times heavier than air and ten times heavier than water; hence, if we divide its unhindered speed into 1000 parts, air will rob it of one of these parts so that it will fall with a speed of 999, while in water its speed will be 900, seeing that water diminishes its weight by one part in ten while air by only one part in a thousand. Again take a solid a little heavier than water, such as oak, a ball of which will weigh let us say 1000 drachms; suppose an equal volume of water to weigh 950, and an equal volume of air, 2; then it is clear that if the unhindered speed of the ball is 1000, its speed in air will be 998, but in water only 50, seeing that the water removes 950 of the 1000 parts which the body weighs, leaving only 50. Such a solid would therefore move almost twenty times as fast in air as in water, since its specific weight exceeds that of water by one part in twenty. And here we must consider the fact that only those substances which have a specific weight greater than water can fall through it—substances which must, therefore, be hundreds of times heavier than air; hence when we try to obtain the ratio of the speed in air to that in water, we may, without appreciable error, assume that air does not, to any considerable extent, diminish the free weight and consequently the speed of such substances. Having thus easily found the excess of the weight of these substances over that of water, we can say that their speed in air is to their speed in water as their net gravity is to the excess of this weight over that of water. For example, a ball of ivory weighs 20 ounces; an equal volume of water weighs 17 ounces; hence the speed of ivory in air bears to its speed in water the approximate ratio of 20 to 3.

Sagr. I have made a great step forward in this truly interesting subject upon which I have long labored in vain. In order to put these theories into practice, we need only discover a method of determining the specific weight of air with reference to water and hence with reference to other heavy substances.

Simp. But if we find that air has levity[13] instead of gravity what then shall we say of the foregoing discussion which, in other respects, is very clever?

Salv. I should say that it was vain and empty. But can you doubt that air has weight when you have the clear testimony of Aristotle affirming that all the elements have weight including air, and excepting only fire? As evidence of this he cites the fact that a leather bottle weighs more when inflated than when collapsed.[14]

Simp. I am inclined to believe that the increase of weight observed in the inflated leather bottle or bladder does not arise from the gravity of the air, but from thick vapors mingled with it lower regions. To this I would attribute the increase of weight in the leather bottle.

Salv. I would not have you say this, and much less attribute it to Aristotle; because, if speaking of the elements, he wished to persuade me by experiment that air has weight and said to me: "Take a leather bottle, fill it with heavy vapors and observe how its weight increases," I would reply that the bottle would weigh still more if filled with bran; and would then add that this merely proves that bran and thick vapors are heavy, but in regard to air I should still remain in the same doubt as before. However, the experiment of Aristotle is good and the proposition is true. But I cannot say as much of a certain other consideration, taken at face value; this consideration was offered by a philosopher whose name slips me; but I know I have read his argument which is that air exhibits greater gravity than levity, because it carries heavy bodies downward more easily than it does light ones upward.

Sagr. Fine indeed! So according to this theory air is much heavier than water, since all heavy bodies are carried downward more easily through air than through water, and all light bodies lifted up more easily through water than through air. In addition, many heavy bodies fall through air but ascend in water, and many substances rise in water and fall in air. But, Simplicio, the question as to whether the weight of the leather bottle comes from thick vapors or from pure air does not affect our problem, which is to discover how bodies move through the atmosphere. Returning now to the question which interests me more, I would like not only to be strengthened in my belief that air has weight but also to learn, if possible, how great its specific weight is. Salviati, please satisfy my curiosity.

Salv. The experiment with the inflated leather bottle of Aristotle proves conclusively that air possesses positive gravity and not, as some have believed, levity, a property possessed possibly by no substance whatever. If air did possess this hypo- *123*
thetical quality, levity should increase under compression and, hence, objects in compressed air should rise more easily; but experiment shows the opposite.

As to the other question, namely, how to determine the specific weight of air, I have employed the following method. I took a rather large glass bottle with a narrow neck and attached to it a leather cover, binding it tightly about the neck of the bottle: in the top of this cover, I inserted and firmly fastened the valve of a leather bottle, through which I forced into the glass bottle, by means/of a syringe, a large quantity of air. Since air is easily compressed one can pump into the bottle two or three times its own volume of air. After this I took an accurate balance and weighed this bottle of compressed air with the utmost precision, adjusting the counterweight with fine sand. I next opened the valve and allowed the compressed air to escape; then replaced the flask upon the balance and found it perceptibly lighter: from the sand which had been used as a counterweight I now removed and laid aside as much as was necessary to again secure balance. Under these conditions there can be no doubt but that the weight of the sand thus laid aside represents the weight of the air which had been forced into the flask and had afterward escaped.

But after all this experiment tells me merely that the weight of the compressed air is the same as that of the sand removed from the balance. When however it comes to knowing the specific weight of air as compared with that of water or any other heavy substance I must first measure the volume of compressed air; for this measurement I have devised the two following methods.

First method: one takes a bottle with a narrow neck similar to the previous one. Over the mouth of this bottle is slipped a leather tube which is bound tightly about the neck of the flask; the other end of this tube embraces the valve attached to the first flask and is tightly bound about it. This second flask is provided with a hole in the bottom through which an iron rod can be placed so as to open, at will, the valve above mentioned and thus permit the surplus air of the first to escape after it has once been weighed: but this second bottle must be filled with water. Having prepared everything in the manner above described, open the valve with the rod; the air will rush into the flask containing the water and will drive it through the hole at the bottom, being clear that the volume of water thus displaced is equal to the volume of air escaped from the other vessel. Having set aside this displaced water, weigh the vessel from which the air has escaped (which is supposed to have been weighed previously while containing the compressed air), and remove the surplus of sand as described above; the weight of this sand is precisely the weight of a volume of air equal to the volume of water displaced and set aside. This water we can weigh and find how many times its weight contains the weight of the removed sand, thus determining how many times heavier water is than air; and we shall find, contrary to the opinion of Aristotle, that this is not 10 times, but, as our experiment shows, more nearly 400 times.

The second method is simpler and can be carried out with a single vessel fitted up as the first was. Here no air is added to that which the vessel naturally contains but water is forced into it without allowing any air to escape; the water thus introduced necessarily compresses the air. Having forced into the vessel as much water as possible, filling it, say, three fourths full, which does not require an extraordinary effort, place it upon the balance and weigh it accurately; next hold the vessel mouth up, open the valve, and allow the air to escape; the volume of the air thus escaping is precisely equal to the volume of water contained in the flask. Again weigh the vessel which will have diminished in weight on account of the escaped air; this loss in weight represents the weight of a volume of air equal to the volume of water contained in the vessel.

Simp. Your devices are clever and smart; but while they appear to give complete intellectual satisfaction, they confuse me. Since it is undoubtedly true that the elements when in their proper places have neither weight nor levity, I cannot understand how it is possible for that portion of air, which appeared to weigh, say, 4 drachms of sand, should really have such a weight in air as the sand which counterbalances it. It seems to me, therefore, that the experiment should be carried out not in air, but in a medium in which air could exhibit its property of weight.

Salv. The objection of Simplicio is certainly appropriate and must therefore either be unanswerable or demand an equally clear solution. It is perfectly evident that that air which, under compression, weighed as much as the sand, loses this weight when once allowed to escape into its own element, while, indeed, sand retains its weight. Hence for this experiment it becomes necessary to select a place where air as well as sand can gravitate, because the medium diminishes the weight of any substance immersed in it by an amount equal to the weight of the displaced medium; so that air in air loses all its weight. If therefore this experiment is to be made with accuracy it should be performed in a vacuum where every heavy body exhibits its momentum without the slightest diminution. If then, Simplicio, we were to weigh a portion of air in vacuum would you then be satisfied and assured of the fact?

Simp. Yes indeed, but this is to ask the impossible.

Salv. I do not want to sell you something which I have already given you: in the previous experiment we weighed the air in vacuum and not in air or other medium. The fact that any fluid medium diminishes the weight of a mass immersed in it, is due to the resistance which this medium offers to its being opened up, driven aside, and finally lifted up. The evidence for this is seen in the readiness with which the fluid rushes to fill up any space formerly occupied by the mass; if the medium were not affected by such an immersion then it would not react against the immersed body. Tell me now: when you have a flask in air, filled with its natural amount of air, and then proceed to pump into the vessel more air, does this extra charge in any way separate or divide or change the external air? Does the vessel perhaps expand so that the surrounding medium is displaced in order to give more room? Certainly not. Therefore we can say that this extra charge of air is not immersed in the surrounding *126* medium, since it occupies no space in it. Indeed, it is really in vacuum: it diffuses into the vacua which are not completely filled by the original and uncondensed air. In fact, I don't see any difference between the enclosed and the surrounding media: the surrounding medium does not press upon the enclosed medium and, vice versa, the enclosed medium exerts no pressure against the surrounding one; this same relationship exists in the case of any matter in vacuum, as well as in the case of the extra charge of air compressed into the flask. The weight of this condensed air is therefore the same as that which it would have if set free in vacuum. It is true of course that the weight of the sand used as a counterweight would be a little greater in vacuum than in free air. We must, then, say that the air is slightly lighter than the sand required to counterbalance it, by an amount equal to the weight in vacuum of a volume of air equal to the volume of the sand.

Sagr.[15] A very clever discussion, solving an interesting problem, because it demonstrates briefly and concisely how one may find the weight of a body in the vacuum by simply weighing it in the air. The explanation is as follows: when a heavy body is immersed in air it loses in weight an amount equal to the weight of a volume of air equivalent to the volume of the body itself. Hence if one adds to a body, without expanding it, a quantity of air equal to that which it displaces, and weighs it, he will obtain its absolute weight in the vacuum, since, without increasing it in size, he has increased its weight by just the amount which it lost through immersion in air. When therefore we force a quantity of water into a vessel that already contains its normal amount of air, without allowing any of this air to escape, it is clear that this normal quantity of air will be compressed and condensed into a smaller space in order to make room for the water which is forced in. It is also clear that the volume of air thus compressed is equal to the volume of water added. If now the vessel is weighed in air in this condition, the weight of the water will be increased by that of an equal volume of air; the total weight of water and air thus obtained is equal to the weight of the water alone in vacuum. Now record the weight of the entire vessel, allow the compressed air to escape, and weigh the remainder. The difference between these two weights will be the weight of the compressed air which, in volume, is equal to that of the water. Next, find the weight of the water alone and add to it that of the compressed air; we shall then have the water alone in the vacuum. To find the weight of the water we shall have to remove it from the vessel and weigh the vessel alone; subtract this weight from that of the vessel and water together. It is clear that the remainder will be the weight of the water alone in air.

Simp. The previous experiments, in my opinion, had left something unexplained: but now I am fully satisfied.

Salv. What I have set forth up to now is new, in particular, the fact that difference of weight, even when very large, does not affect the speed of falling bodies. This idea is at first glance so counterintuitive, that if we did not have the means of making it just as clear as sunlight, it had better not be mentioned; but having once allowed it to pass my lips I must not neglect any experiment or argument to establish it.

Sagr. Not only this but also many other of your views are so far from the commonly accepted opinions and doctrines that if you were to publish them you would stir up a large number of antagonists, since human nature is such that people do not look with favor new discoveries in their own field when made by others than themselves. They hope to cut those knots which they cannot untie, and they try to destroy structures that patient artisans have built. But we have no such thoughts: experiments and arguments which you have thus far adduced are fully satisfactory. However, if you have any more direct experiments, or any arguments which are even more convincing, we will hear them with pleasure.

Salv. The experiment made to ascertain whether two bodies, differing greatly in weight, will fall from a given height with the same speed, offers some difficulty; because, if the height is considerable, the retarding effect of the medium, which must be penetrated and thrust aside by the falling body, will be greater in the case of the small momentum of the very light body than in the case of the great weight of the heavy body; so that, in a long distance, the light body will be left behind. If instead the height is small, one may well doubt whether there is any difference; and if there be a difference it will be inappreciable.

It occurred to me therefore to repeat many times the fall through a small height in such a way that I might accumulate all those small intervals of time between the arrival of the heavy and light bodies respectively at their common terminus, so that this sum makes an interval of time which is not only observable, but easily observable. To employ the slowest possible speeds and thus reduce the change that the resisting medium produces upon the simple effect of gravity I made the bodies fall along a plane slightly inclined to the horizontal. On such a plane, just as well as on a vertical plane, one may discover how bodies of different weight behave. Besides this, I also wished to get rid of the resistance which might arise from contact of the moving body with the aforesaid inclined plane. Accordingly, I took two balls, one of lead and one of cork, the former more than a hundred times heavier than the latter, and suspended them by means of two equal fine threads, each four or five cubits long. Pulling each ball aside from the perpendicular, I let them go at the same moment, and they, falling along the circumferences having these equal ropes as radii, passed beyond the perpendicular and returned along the same path. This oscillation repeated a hundred times showed clearly that the heavy body maintains so nearly the period *129* of the light body that neither in a hundred oscillations nor even in a thousand will the former anticipate the latter in an appreciable way. We can also observe the effect of the medium which, by the resistance which it offers to motion, diminishes the vibration of the cork more than that of the lead, but without altering the period of either; even when the arc traversed by the cork did not exceed five or six degrees while that of the lead was fifty or sixty, the oscillations were performed in equal times.

Simp. If so, why is not the speed of the lead greater than that of the cork, seeing that the former traverses sixty degrees in the same interval in which the latter covers scarcely six?

Salv. What would you say, Simplicio, if both covered their paths in the same time when the cork, drawn aside through thirty degrees, traverses an arc of sixty, while the lead pulled aside only two degrees traverses an arc of four? Would not then the cork be proportionately swifter? And yet such is the experimental fact. But observe this: having pulled aside the pendulum of lead, say through an arc of fifty degrees, and set it free, it swings beyond the perpendicular almost fifty degrees, thus describing an arc of nearly one hundred degrees; on the return, it describes a little smaller arc; and after a large number of such vibrations it finally comes to rest. Each vibration, whether of ninety, fifty, twenty, ten, or four degrees takes the same time: accordingly the speed of the moving body keeps on diminishing since in equal intervals of time, it traverses smaller and smaller arcs. The same things happen with the pendulum of cork, suspended by a cord of equal length, except that a smaller number of vibrations is required to bring it to rest, since on account of its lightness it is less able to overcome the resistance of the air; nevertheless the oscillations, whether large or small, are all performed in time intervals which are not only equal among themselves, but also equal to the period of the lead pendulum. Hence it is true that, if while the lead is traversing an arc of fifty degrees the cork covers one of only ten, the cork moves more slowly than the lead; but on the other hand it is also true that the cork may *130* cover an arc of fifty degrees while the lead passes over one of only ten or six; thus, at different times, we have now the cork, now the lead, moving more rapidly. But if these same bodies traverse equal arcs in equal times we may rest assured that their speeds are equal.

Simp. I hesitate to admit the conclusiveness of this argument because of the confusion which arises from your making both bodies move now rapidly, now slowly and now very slowly, which leaves me in doubt as to whether their velocities are always equal.

Sagr. Allow me, if you please, to say just a few words. Now tell me, Simplicio: do you admit that one can be sure that the speeds of the cork and the lead are equal whenever both, starting from rest at the same time and descending the same slopes, always traverse equal spaces in equal times?

Simp. This can neither be doubted nor contradicted.

Sagr. Now it happens, in the case of pendulums, that each of them traverses now an arc of sixty degrees, now one of fifty, or thirty or ten or eight or four or two, etc.; and when they both swing through an arc of sixty degrees they do so in equal intervals of time. The same thing happens when the arc is fifty degrees or thirty or ten or any other number; and therefore we conclude that the speed of the lead in an arc of sixty degrees is equal to the speed of the cork when the latter also swings through an arc of sixty degrees. In the case of a fifty-degree arc these speeds are also equal to each other; so also in the case of other arcs. But this is not saying that the speed which occurs in an arc of sixty is the same as that occurring in an arc of fifty; nor is the speed in an arc of fifty equal to that in one of thirty, etc.; but the smaller the arcs, the smaller the speeds. The fact observed is that one and the same moving body requires the same time for traversing a large arc of sixty degrees as for a small arc of fifty or even a very small arc of ten degrees; all these arcs, indeed, are covered in 131 the same interval of time. It is true therefore that the lead and the cork each diminish their speed in proportion as their arcs diminish; but this does not contradict the fact that they maintain equal speeds in equal arcs. My reason for saying these things has been rather because I wanted to learn whether I had correctly understood Salviati, than because I thought Simplicio had any need of a clearer explanation than that given by Salviati. His explanations are always extremely lucid, so lucid that when he solves questions which are difficult not merely in appearance, but in reality and in fact, he does so with reasons, observations and experiments which are common and familiar to everyone. In this manner he has given occasion to a highly esteemed professor for undervaluing his discoveries on the ground that they are commonplace, and established upon a mean and vulgar basis; as if it were not a most admirable and praiseworthy feature of demonstrative science that it springs from and grows out of well-known principles, understood and conceded by all. But let us continue with this light diet; and if Simplicio is satisfied to understand and admit that the gravity inherent in various falling bodies has nothing to do with the difference of speed observed among them, and that all bodies, in so far as their speeds depend upon it, would move with the same velocity, please tell us, Salviati, how you explain the appreciable and evident inequality of motion. Please reply also to the objection urged by Simplicio—an objection in which I concur– namely, that a cannonball falls more rapidly than a bird-shot. From my point of view, one might expect the difference of speed to be small in the case of bodies of the same substance moving through any single medium, whereas the larger ones will descend, during a single pulse-beat, a distance which the smaller ones will not traverse in an hour, or four, or even in twenty hours; for instance, in the case of stones and fine sand and especially that very fine sand which produces muddy water and which in many hours will not fall through as much as two cubits, a distance that stones not much larger will traverse in a single pulse-beat.

Salv. The action of the medium in producing greater retardation on bodies that have a smaller specific weight has already been explained by showing that they 132 experience a diminution of weight. But to explain how one and the same medium produces such different retardations in bodies made of the same material and with the same shape, but different only in size, requires a deeper discussion.

The solution to this problem lies, I think, in the roughness and porosity which are generally and almost necessarily found on the surfaces of solid bodies. When the body is in motion these roughnesses strike the air or other ambient medium. The evidence for this is found in the humming which accompanies the rapid motion of a body through air, even when that body is as round as possible. One hears not only humming, but also hissing and whistling, whenever there is any appreciable cavity or elevation upon the body. We observe also that a rotating round solid body produces a current of air. But what more do we need? When a top spins on the ground at its greatest speed don't we hear a distinct buzzing of high pitch? This sibilant note diminishes in pitch as the speed of rotation decreases, which is evidence that these small rugosities on the surface meet resistance in the air. There can be no doubt, therefore, that in the motion of falling bodies these rugosities strike the surrounding fluid and retard the speed of the moving body; and this they do so much the more the surface is larger in proportion, which is the case of smaller bodies as compared with greater bodies.

Simp. Stop a moment, please: I am getting confused. Although I understand and admit that friction of the medium upon the surface of the body retards its motion and that, if other things are the same, the larger surface suffers greater retardation, I do not see on what ground you say that the surface of the smaller body is larger. Besides if, as you say, the larger surface suffers greater retardation, the larger solid should move more slowly, which is not the fact. But this objection can be easily met by saying that, although the larger body has a larger surface, it has also a greater weight, in comparison with which the resistance of the larger surface is no more than the resistance of the small surface in comparison with its smaller weight; so that the speed of the larger solid does not become less. I therefore see no reason for expecting any difference of speed so long as the driving weight diminishes in the *133* same proportion as the retarding power of the surface.

Salv. I shall answer all your objections at once. You will admit, Simplicio, that if one takes two equal bodies of the same material and same shape, bodies which would therefore fall with equal speeds, and if he diminishes the weight of one of them in the same proportion as its surface (maintaining the similarity of shape) he would not thereby diminish the speed of this body.

Simp. It should be so, if I accept that the weight of a body has no effect in either accelerating or retarding its motion.

Salv. I quite agree with you in this opinion. It follows that, if the weight of a body is diminished in greater proportion than its surface, the motion is retarded to a certain extent; and this retardation is greater and greater in proportion as the diminution of weight exceeds that of the surface.

Simp. This I admit without hesitation.

Salv. Now you must know, Simplicio, that it is not possible to diminish the surface of a solid body in the same ratio as its weight, and at the same time maintain the similarity of shape. For since it is clear that in the case of a diminishing solid the weight grows less in proportion to the volume, and since the volume always diminishes more rapidly than the surface, when the same shape is maintained, the weight must therefore diminish more rapidly than the surface. But geometry teaches us that, in the case of similar solids, the ratio of two volumes is greater than the ratio of their surfaces; which, for the sake of better understanding, I shall illustrate by a particular case. Take, for example, a cube two inches on a side so that each face has an area of four square inches and the total area, i.e., the sum of the six faces, amounts to twenty-four square inches; now imagine this cube to be sawed through three times so as to divide it into eight smaller cubes, each one inch on the side, each face one inch square, and the total surface of each cube is six square inches instead of twenty-four as in the case of the larger cube. It is evident therefore that the surface of the little cube is only one-fourth that of the larger, namely, the ratio of six to twenty-four; but the volume of the solid cube itself is only one eighth; the volume, and hence also the weight, diminishes therefore much more rapidly than the surface. If we again divide the little cube into eight others we shall have, for the total surface of one of these, one and one-half square inches, which is one sixteenth of the surface of the original cube; but its volume is only one-sixty-fourth part. Thus, by two divisions, you see that the volume is diminished four times as much as the surface. And, if the subdivision is continued until the original solid be reduced to a fine powder, we shall find that the weight of one of these smallest particles has diminished hundreds and hundreds of times as much as its surface. And this which I have illustrated in the case of cubes holds also in the case of all similar solids. Given a linear dimension of a body, L, calling V its volume and S its surface, one has

$$S \propto L^2 \; ; \; V \propto L^3$$

and thus

$$\frac{V}{S} \propto L .$$

Observe then how much greater the resistance, arising from contact of the surface of the moving body with the medium, in the case of small bodies than in the case of large ones, and when one considers that the rugosities on the very small surfaces of fine dust particles are perhaps no smaller than those on the surfaces of larger solids which have been carefully polished, he will see how important it is that the medium is very fluid and offers no resistance to being thrust aside, easily yielding to a small force. You see, therefore, Simplicio, that I was not mistaken when I said that the surface of a small solid is comparatively greater than that of a large one.

Simp. I am quite convinced; and, believe me, if I were again beginning my studies, I would follow the advice of Plato and start with mathematics, a science that proceeds very cautiously and admits nothing as established until it has been rigidly demonstrated.

Sagr. I appreciate what you said. Before going ahead, Salviati, I would like to understand what did you mean when you said that the ratios between volumes and surfaces of similar solids stand between them as the linear dimension.

Salv. I mean that the volumes (and thus the weights) grow proportionally to the cubes of the linear dimensions, and surfaces (and thus resistances) proportionally to the squares.

Sagr. I understood, and I liked this reasoning very much. There are some details that I'd like to be clarified, but I'm afraid that if we continue from one digression to another we will miss our initial problem, namely the different aspects of the resistance of solids to be broken. Please tie again the threads.

Salv. Very well; but the questions which we have already considered are so numerous and so varied, and have taken up so much time that there is not much of this day left to spend upon our main topic which abounds in geometrical demonstrations calling for careful consideration. May I, therefore, suggest that we postpone the meeting until tomorrow, not only for the reason just mentioned but also in order that I may bring with me some papers in which I have set down in an orderly way the theorems and propositions dealing with the various phases of this subject, matters which, from memory alone, I could not present in the proper order.

Sagr. I fully agree, also because this will leave time today to take up some of my difficulties with the subject which we have just been discussing. One question is whether we are to consider the resistance of the medium as sufficient to destroy the acceleration of a body of very heavy material, very large volume, and spherical figure. I say spherical to select a volume that is contained within a minimum surface and therefore less subject to retardation. Another question deals with the vibrations of pendulums which may be regarded from several viewpoints; the first is whether all vibrations, large, medium, and small, are performed in exactly and precisely equal times: another is to find the ratio of the times of vibration of pendulums supported by threads of unequal length. *136*

Salv. These are interesting questions: but I fear that here, as in the case of all other facts, if we take up for discussion any one of them, it will carry in its wake so many other facts and curious consequences that time will not remain today for the discussion of all.

Sagr. If these are as full of interest as the foregoing, I would gladly spend as many days as there remain hours between now and nightfall; and I dare say that Simplicio would not be annoyed by these discussions.

Simp. Certainly not; especially when the questions pertain to physics and have not been treated by other philosophers.

Salv. Now taking up the first question, I can assert without hesitation that there is no sphere so large, or composed of material so dense, that the resistance of the medium, although very slight, would not contrast its acceleration and after a certain time reduce its motion to uniformity; a statement which is strongly supported by experiment. If a falling body, as time goes on, were to acquire a speed as great as you please falling through a medium, then no speed that could be conferred on it by a mover external to that medium could be so great that the moveable would reject it due to the impediment of that medium. Thus, for instance, if a cannonball, having fallen a distance of four cubits through air and having acquired a speed of, say, ten units, were to strike the surface of the water, and if the resistance of the water were not able to contrast the impetus of the shot, it would either increase in speed or maintain a uniform motion until the bottom is reached. But this is not what we observe; on the contrary, water only a few cubits deep slows down the motion in such a way that the shot hits the bed of the river or lake with a very light impact. Clearly then if a short *137*

fall through water is sufficient to deprive a cannonball of its speed, this speed cannot
be regained by a fall of even a thousand cubits. How could a body acquire, in a fall
of a thousand cubits, the speed which it loses in a fall of four? Do we need more?
Don't we observe that the enormous impetus delivered to a shot by a cannon is so
weakened by passing through a few cubits of water that the ball, rather than injuring
the ship, hardly strikes it?

Air can also diminish the speed of a falling body, as may be easily understood
from similar experiments. If a gun fires downwards from the top of a very high tower
the shot will have a smaller effect on the ground than if the gun had been fired from
an elevation of only four or six cubits; this is clear evidence that the momentum
of the ball, fired from the top of the tower, diminishes in descending through air.
Therefore a fall from ever so great an altitude will not suffice to give to a body that
momentum which it has once lost through the resistance of the air, no matter how
it was originally acquired. Similarly, the damage produced upon a wall by a shot
fired from a gun at a distance of twenty cubits cannot be duplicated by the fall of
the same shot from any altitude however great. My opinion is, therefore, that under
the circumstances which occur in nature, the acceleration of any body falling from
rest reaches a maximum value and that the resistance of the medium finally brings
its speed to a constant value which is thereafter maintained.

Sagr. These experiments are in my opinion much to the purpose. The only question
is whether this is still true in the case of bodies that are very large and heavy or very far
away: a cannonball, falling from the distance of the Moon or from the upper regions
of the atmosphere, would deliver a heavier blow than if just leaving the muzzle of
the gun?

Salv. No doubt many objections may be raised, not all of which can be refuted by
experiment. It is very likely that a heavy body falling from a height will, on reaching
the ground, have acquired just as much impetus as it is necessary to carry it to that
height; as may be clearly seen in the case of a rather heavy pendulum which, when
pulled aside fifty or sixty degrees from the vertical, will acquire precisely that speed
which is sufficient to carry it to an equal elevation, apart only that small portion which
it loses through friction. To place a cannonball at such a height as might suffice to
give it just that momentum which the powder imparted to it on leaving the gun we
need only to fire it vertically upwards from the same gun; and we can then observe
whether on falling back it delivers a blow equal to that of the gun fired at close
range—in my opinion it would be much weaker. The resistance of the air would
therefore, I think, prevent the velocity from being equaled by a natural fall from rest
at any height whatever.

We come now to the other questions related to pendulums, a subject which may
appear arid, especially to those philosophers who are continually occupied with
the more profound questions of nature. Nevertheless, the problem is one that I do
not scorn. I am encouraged by the example of Aristotle whom I admire especially
because he did not fail to discuss every subject which he thought to any degree
worthy of consideration. Impelled by your questions I may give you some of my
ideas concerning certain problems in music, a splendid subject, upon which so many
eminent people have written—among these is Aristotle himself, who has discussed

138-
140

numerous interesting questions related to acoustics. Accordingly, on the basis of easy
and tangible experiments I shall explain some striking phenomena in the domain of
sound, and I trust my explanations will be welcome to you.

Sagr. I shall receive them not only gratefully but eagerly. For, although I take
pleasure in every kind of musical instrument and have paid considerable attention
to harmony, I have never been able to fully understand why some combinations of
tones are more pleasing than others, or why some others not only fail to please but
are even highly offensive. Then there is the old problem of two stretched strings in
unison; when one of them is sounded, the other begins to vibrate and to emit its note
Nor do I understand the different ratios of harmony and some other details.

Salv. Let's see whether we can derive from the pendulum a satisfactory solution
to all these difficulties.

And first, as to the question of whether a pendulum really performs its vibrations
(large, medium, and small) all in exactly the same time, I shall rely upon what I
have already heard from our Academician. He has clearly shown that the time of
descent is the same along all chords, whatever the arcs which subtend them, as well
along an arc of 180 degrees (i.e., the full diameter) as along one of fractions of a
degree, as long as they terminate at the lowest point of the circle, where it touches the
horizontal plane.[1] If now we consider the descent along arcs instead of their chords
then, provided these do not exceed 90 degrees, experiment shows that they are all
traversed in equal times; but these times are greater for the chord than for the arc,
an effect which is all the more remarkable because at first glance one would think
just the opposite to be true. For since the terminal points of the two motions are the
same and since the straight line included between these two points is the shortest
distance between them, it would seem reasonable that motion along this line should
be executed in the shortest time; but this is not the case, as the shortest time—and
therefore the most rapid motion—is that employed along the arc of which this straight
line is the chord.

As for the periods T of the oscillations of bodies hanging from strings of different
lengths, we observe that they are proportional to the square root of the lengths L of
the strings.

Sagr. Then, if I understand correctly, I can calculate the length of a string hanging
from a great height, even if the upper end is invisible to me and only the other extreme
can be seen. If I attach a weight to the lower end of the string and I let it oscillate,
and ask a friend to count the number of oscillations while in the same time interval
I count the oscillations of another body hanging from a thread of the length of one
cubit, from the comparison of the numbers of oscillations of these pendulums in the
same time interval I will find the length of the cord:

$$T \propto \sqrt{L} \; ; \; L \propto T^2 . \tag{6}$$

Suppose, for example, that my friend counts 20 oscillations of the long string during
the same time in which I count 240 oscillations of my string which is one cubit in

---

[1] This proof will be presented on the third day.

length. The ratio between the periods is 12, and thus the long string is 12^2 times 1 cubit = 144 cubits.

Salv. You can make this measurement very accurate, especially if you measure a large number of oscillations.

Sagr. You often give me occasion to admire the wealth and profusion of nature when, from such common and even trivial phenomena, you derive facts which are not only striking and new but which are often far removed from what we would have imagined. Thousands of times I have observed vibrations especially in churches where lamps, suspended by long cords, had been inadvertently set into motion; but the most which I could infer from these observations was that the view of those who think that such vibrations are maintained by the medium, i.e., air, is highly improbable. Air should have considerable judgment and little else to do in order to spend hours and hours pushing back and forth a hanging weight with perfect regularity! And I never dreamed of learning that a body, when suspended from a string a hundred cubits long and pulled aside through an arc of 90 degrees or even one degree or a fraction of a degree, would employ the same time in passing through these arcs—indeed, it still strikes me as somewhat unlikely. Now I am waiting to hear how these same simple phenomena can provide solutions, at least partially, for acoustical problems.

Salv. First of all, one must observe that each pendulum has its own time of oscillation, so definite and determinate that it is not possible to make it move with any other period than that which nature has given it. If we take a pendulum and try to increase or diminish the frequency of its vibrations, we will waste our time. On the other hand, one can confer motion even to a heavy pendulum at rest by simply blowing against it; by repeating these blasts with a frequency that is the same as that of the pendulum, one can impart considerable motion. Suppose that by the first puff we have displaced the pendulum from the vertical by, say, half an inch; then if, after the pendulum has returned and is about to begin the second vibration, we add a second puff, we shall impart additional motion; and so on with other blasts provided they are applied at the right time, and not when the pendulum is coming toward us since in this case the blast would impede rather than aid the motion. Continuing thus with many impulses we impart to the pendulum such an impetus that a much greater force than that of a single blast will be needed to stop it.

Sagr. When I was a boy, I observed that one man alone by giving impulses at the right moment was able to ring a bell so large that when four, or even six, men seized the rope and tried to stop it, they were lifted from the ground, all of them together being unable to counterbalance the momentum which a single man, by properly timed pulls, had given it.

Salv. Your illustration makes my meaning clear and is quite as well fit to explain the wonderful phenomenon of the strings of the cittern or of the spinet, namely, the fact that a vibrating string will set another string in motion and cause it to sound not only when the latter is in unison (same musical note), but even when it differs from the former by an octave or a fifth.[m] A string which has been struck begins to vibrate and continues the motion as long as one hears the sound; these vibrations cause the air immediately surrounding to vibrate, and ripples in the air expand far into space and strike not only all the strings of the same instrument but even those of neighboring instruments. Since that string which is tuned to unison with the one plucked is capable of vibrating with the same frequency, it acquires, at the first impulse, a slight oscillation; after receiving two, three, twenty, or more impulses, since these are by construction delivered at proper intervals, it finally accumulates a vibratory motion equal to that of the plucked string, as is clearly shown by equality of amplitude in their vibrations.

Wave action expands through the air and sets into vibration not only strings, but also any other body which has the same period as the plucked string. Accordingly, if we attach to the side of an instrument small pieces of bristle or other flexible material, we shall observe that, when a spinet is played, only those pieces respond that have the same period as the string which has been struck; the remaining pieces do not vibrate in response to this string, nor do the former pieces respond to different strings. If one bows strongly a thick string on a viola and brings near it a fine goblet, the glass having the same tone as that of the string, this goblet will vibrate and audibly resound. That the undulations of the medium are widely dispersed about the sounding body is evinced by the fact that a glass of water may be made to emit a tone merely by the friction of the finger-tip upon the rim of the glass, and a series of regular waves are produced in the water. The same phenomenon is observed by fixing the base of the goblet upon the bottom of a rather large vessel of water filled near to the edge of the goblet; if, as before, we sound the glass by the friction of the finger, we shall see ripples spreading with the utmost regularity and with high speed to large distances about the glass. I have often remarked, in sounding a rather large *143* glass nearly full of water, that at first the waves are spaced with great uniformity, and when, as sometimes happens, the tone of the glass jumps an octave higher, at this moment each of the aforesaid waves divides into two; a phenomenon which shows clearly that the ratio of frequencies and wavelengths involved in the octave is two.

Sagr. More than once have I observed this same thing, much to my delight and also to my profit. For a long time I have been perplexed about these different harmonies since the explanations hitherto given by those learned in music. i.e., that these ratios on the lengths of the strings are simple and natural, seem to me not sufficiently conclusive.[n]

My idea is the following. There are three ways to make the pitch of a string more acute: one is to shorten it; another is to stretch it more; the third is to make it thinner. If we keep the same tension and the same thickness of the string, and we want to *144*

---

[m] Octave: the string has a double (upper octave) or half (lower octave) frequency with respect to the first one; a higher octave can be achieved for example by blocking the initial string in the middle and swinging one of the two halves. Fifth interval (higher): the second string has a frequency equal to 3/2 of the first; an upper fifth can be realized for example by locking the initial string to a third and swinging the shorter part.

[n] He is also quoting his father, in his treatise *Dialogo della musica antica et della moderna* (1581), p. 13.

hear the octave, we must shorten it by one half; that is, pinch it, put a bridge in the middle and then pinch it again. Keeping the same length and thickness, if we want to raise the sound of an octave we will need a tension T four times larger; so if initially the string was kept in tension by the weight of one pound, it will be necessary to attack four pounds to it. Finally, if we keep the same length and tension, and we want a string that sounds an octave higher, we will have to reduce the section Σ to one quarter. In short, the fundamental frequency of oscillation ν will be, for a given material,

$$\nu \propto \frac{1}{\ell}\sqrt{\frac{T}{\Sigma}}. \tag{7}$$

This relationship applies to all consonances. If we consider the fifth, it can be obtained again as well by changing the tension or the thickness. So if the rope is stretched by a weight of four pounds, to obtain the higher pitch it will be necessary to attack nine; and as regards the thickness, to obtain the fifth, the section of the string must be larger by a ratio of nine to four.

Since it is impossible to count the vibrations of a sounding string on account of its high frequency, I should still have been in doubt as to whether a string, emitting the upper octave, made twice as many vibrations in the same time as one giving the fundamental, had it not been for the fact that at the moment when the tone jumps to the octave, the waves which constantly accompany the vibrating glass divide up into smaller ones which are precisely half as long as the former.

Salv. A beautiful experiment, enabling us to distinguish individually the waves produced by the vibrations of a resounding body. Such waves spread through the air, bringing to the eardrum a stimulus which mind translates into sound. But since these waves in the water last only as long as the friction of the finger continues and are, even then, not constant but always forming and disappearing, would it not be a fine thing if one could produce waves persisting for a long while, even months and years, so as to easily measure and count them?

Sagr. Such an invention would, I assure you, excite my admiration.

Salv. The device is one which I hit upon by accident; my part consists merely in the observation of it and in the appreciation of its value as a confirmation of something to which I had given profound consideration; and yet the device is, in itself, rather common. As I was scraping a brass plate with a sharp iron chisel to remove some spots from it and was running the chisel rather rapidly over it, I once or twice, during many strokes, heard the plate emit a rather strong and clear whistling sound; on looking at the plate more carefully, I noticed a long row of fine streaks parallel and equidistant from one another. Scraping with the chisel over and over again, I noticed that it was only when the plate emitted this hissing noise that any marks were left upon it; when the scraping was not accompanied by this sibilant note there was not the least trace of such marks. Repeating the trick several times and making the stroke, now with greater now with less speed, the whistling followed with a pitch which was correspondingly higher and lower. I noted also that the marks made when the tones were higher were closer together; but when the tones were deeper, they were farther apart. I also observed that when, during a single stroke, the speed increased toward the end, the sound became sharper and the streaks closer, but always in such a way as to remain sharply defined and equidistant. Whenever the stroke was sibilant I felt the chisel tremble in my hand and a sort of shiver run through my arm. In short, we see and hear in the case of the chisel what is seen and heard in the case of a whisper followed by a loud voice: when the breath is emitted without the production of a tone, one does not feel either in the throat or mouth any motion to speak of in comparison with that which is felt in the larynx and upper part of the throat when the voice is used, especially when the tones employed are low and strong.

Sometimes I have also observed among the strings of the spinet two which were in unison with two of the tones produced by the aforesaid scraping; and among those which differed most in pitch, I found two which were separated by an interval of a perfect fifth. Measuring the distance between the marks produced by the two scrapings I found that the space which contained 45 of one contained 30 of the other, which is precisely the ratio assigned to the fifth.

146-147

But now before proceeding any farther I want to call your attention to the fact that, of the three methods for sharpening a tone, the one which you refer to as the fineness of the string could be attributed to its weight. As long as the material of the string is unchanged, the size and weight vary in the same ratio. Thus in the case of gut strings, we obtain the octave by making one string 4 times as large as the other; so also in the case of brass one wire must have 4 times the size of the other; but if now we wish to obtain the octave of a gut string, by use of brass wire, we must make it, not four times as large, but four times as heavy as the gut string: as regards size, therefore, the metal string is not four times as big but four times as heavy. In summary

$$\nu \propto \frac{1}{\ell}\sqrt{\frac{T}{\mu}}, \tag{8}$$

where μ is the weight per unit length:

$$\mu = \frac{w}{\ell} = \frac{w\Sigma}{\ell\Sigma} = \frac{w\Sigma}{V} = \gamma\Sigma$$

(I called γ the specific weight of the material).

A brass string may therefore be thinner than a gut string even though the latter gives the higher note. Hence if two spinets are strung, one with gold wire the other with brass, and if the corresponding strings each have the same length, diameter, and tension, it follows that the instrument strung with gold will have a pitch about one fifth lower than the other because gold has a density almost twice that of brass. And here it is to be noted that it is the weight rather than the size of a moving body that offers resistance to change of motion contrary to what one might think at first glance. For it seems reasonable to believe that a body that is large and light should suffer greater retardation of motion in thrusting aside the medium than would one which is thin and heavy; yet here exactly the opposite is true. Returning now to the original subject of discussion, I assert that the ratio of a musical interval is not immediately determined either by the length, size, or tension of the strings but rather by the ratio of their frequencies, that is, by the number of pulses of air waves which strike the eardrum, causing it also to vibrate with the same frequency. This fact established, we may possibly explain why certain pairs of notes differing in pitch produce a pleasing sensation, others a less pleasant effect, and still others a disagreeable sensation. Such an explanation could lead to an explanation of the more or less perfect consonances and of dissonances. The unpleasant sensation produced by the latter arises, I think, from the discordant vibrations of two different tones which strike the ear out of time. Especially harsh is the dissonance between notes whose frequencies are incommensurable; such a case occurs when one has two strings in unison and sounds one of them open, together with a part of the other which is shorter than the first by a factor $\sqrt{2}$.[°]

Agreeable consonances are pairs of tones striking the ear with a certain regularity; this regularity consists in the fact that the pulses delivered by the two tones, in the same interval of time, shall be commensurable in number, so as not to keep the ear drum in perpetual torment, bending in two different directions to yield to the ever discordant impulses. The first and most pleasing consonance is, therefore, the octave since, for every pulse given to the eardrum by the lower string, the sharp string delivers two; accordingly, at every other vibration of the upper string, both pulses are delivered simultaneously so that one half the entire number of pulses are delivered in unison. But when two strings are in unison their vibrations always coincide and the effect is that of a single string; hence we do not refer to it as consonance. The fifth is also a pleasing interval since for every two vibrations of the lower string the upper one gives three, so that considering the entire number of pulses from the upper string one-third of them will strike in unison, i.e., between each pair of concordant vibrations there intervene two single vibrations; and when the interval is a fourth, three single vibrations intervene. In case the ratio is 9/8, only every ninth vibration of the upper string reaches the ear simultaneously with one of the lower; all the others are discordant and produce a harsh effect upon the recipient ear which interprets them as dissonances.

Simp. Could you please explain this argument a little more clearly?

[°] On the tempered sequence (e.g., the sequence of the piano keys) we produce this dissonance by playing two keys six keys apart, without distinguishing between white and black keys—that is, by dividing an octave into two harmonically equal parts. This dissonance is called tritone, and it was absolutely to be avoided according to medieval harmony theorists like Guido d' Arezzo—who called it "the devil". Today we are more tolerant, and we listen to it without too much disturbance, for example in the aria *Mariah* from *West side story* by Bernstein, in the theme of *the Simpsons* and from the sirens of the ambulances.

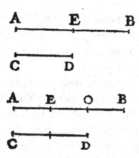

Salv. Let AB denote the length of vibration by the lower pitch string and CD that of a higher pitch string which is emitting the octave of AB; divide AB in the center at E. If the two strings begin their motions at A and C, it is clear that when the sharp vibration has reached the end D, the other vibration will have traveled only as far as E, which, not being a terminal point, will emit no pulse; but there is a blow delivered at D. When the first wave comes back from D to C, the other passes on from E to B; hence the two pulses from B and C strike the eardrum simultaneously. Seeing that these vibrations are repeated again and again, we conclude that each alternate pulse from CD falls in unison with one from AB.

Next, let the vibrations AB and CD be separated by an interval of a fifth, that is, be in the ratio of 3/2; choose the points E and O such that they divide the wavelength of the lower string into three equal parts and imagine the vibrations to start at the same moment from each of the terminals A and C. It is evident that when the pulse has been delivered at the terminal D, the wave in AB has traveled only as far as O; the eardrum receives, therefore, only the pulse from D. Then during the return of the first vibration from D to C, the other will pass from O to B and then back to O, producing an isolated pulse at B—a pulse which is out of time but one which must be taken into consideration. Now since we have assumed that the first pulsations started from the terminals A and C at the same moment, it follows that the second pulsation, isolated at D, occurred after an interval of time equal to that required for passage from C to D or, what is the same thing, from A to O; but the next pulsation, the one at B, is separated from the preceding by only half this interval, namely, the time required for passage from O to B. Next while the one vibration travels from O to A, the other travels from C to D, the result of which is that two pulsations occur simultaneously at A and D. Cycles of this kind follow one after another, i.e., one solitary pulse of the lower string interposed between two solitary pulses of the upper string. Let us now imagine time to be divided into very small equal intervals; then we assume that, during the first two of these intervals, the pulsations which occurred simultaneously at A and C have traveled as far as O and D and have produced a pulse at D; and that during the third and fourth intervals one vibration returns from D to C, producing a pulse at C, while the other, passing on from O to B and back to O, produces a pulse at B; and finally, during the fifth and sixth intervals, the vibrations travel from O and C to A and D, producing a pulse at each of the latter two. Then the sequence in which the pulses strike the ear will be such that, if we begin to count time from any moment where two pulses are simultaneous, the ear drum will, after the lapse of two of the

148

said intervals, receive a solitary pulse; and at the end of the third interval, another solitary pulse. So also at the end of the fourth interval; and two intervals later, i.e., at the end of the sixth interval, will be heard two pulses in unison. Here ends the cycle, which repeats itself over and over again.

Sagr. I had great pleasure in hearing such a complete explanation of phenomena concerning which I have so long been in darkness. Now I understand why unison does not differ from a single tone; I understand why the octave is the principal harmony, and, like unison, it can mix with other consonances. It resembles unison because the pulsations of strings in unison always occur simultaneously, and those of the lower string of the octave are always accompanied by those of the upper string; and among the latter a solitary pulse is interposed at equal intervals and in such a manner as to produce no disturbance. The result is that such a harmony is rather too softened and lacks fire. But the fifth is characterized by its displaced beats and by the interposition of two solitary beats of the upper tone and one solitary beat of the lower tone between each pair of simultaneous pulses; these three solitary pulses are separated by intervals of time equal to half the interval which separates each pair of simultaneous beats from the solitary beats of the upper string. Thus the effect of the fifth is to produce a tickling of the ear drum such that its softness is modified with sprightliness, giving simultaneously the impression of a gentle kiss and of a bite.

Salv. Seeing that you have derived so much pleasure from these novelties, I must show you a method by which the eye may enjoy the same game as the ear. Suspend three balls of lead, or other heavy material, by means of strings of different length such that while the longest makes two vibrations the shortest will make four and the medium three; this will take place when the longest string measures 16, either spans or any other unit, the medium 9, and the shortest 4, all measured in the same unit. Now pull all these pendulums aside from the perpendicular and release them at the same moment; you will see a curious interplay of the threads passing each other in various manners but such that at the completion of every fourth vibration of the longest pendulum, all three will arrive simultaneously at the same terminus, whence they start over again to repeat the same cycle. This combination of vibrations is precisely that which yields the interval of the octave and the intermediate fifth. If we employ the same disposition of apparatus but change the lengths of the threads, always however in such a way that their vibrations correspond to those of agreeable musical intervals, we shall see a different crossing of these threads but always such that, after a definite interval of time and after a definite number of vibrations, all the threads, whether three or four, will reach the same point at the same moment, and then begin a repetition of the cycle. If, however, the vibrations of two or more strings are incommensurable so that they never complete a definite number of vibrations at the same moment, or if commensurable they return only after a long interval of time and after a large number of vibrations, then the eye is confused by the disordered succession of crossed threads. In like manner, the ear is pained by an irregular sequence of air waves that strike the eardrum without any fixed order.

But where have we drifted during these many hours by various problems and unexpected digressions? The day is already ended and we have scarcely touched the main subject we wanted to discuss. Indeed we have deviated so far that I remember only with difficulty our early introduction and the little progress made in the way of hypotheses and principles for use in later demonstrations.

Sagr. Let's then adjourn for today so that our minds may find refreshment in sleep and that we may return tomorrow, if so please you, and resume the discussion of the main question.

Salv. I shall not fail to be here tomorrow at the same hour, hoping not only to render you service but also to enjoy your company.

The first day ends.

Day Two

(What Could Be the Cause of Cohesion)

Sagredo. Waiting for your arrival, Simplicio and I were remembering the considerations of yesterday, which will serve as a basis for what you want to show us about the
resistance of solid bodies to fracture. The strength of bodies depends on the cement
that holds them together and attaches their parts, in such a way that only a strong pull
can separate them. After searching what could be the cause of this cohesion, which
in some solids is very strong, and examining if this cause could be the fear of the
vacuum, we have started many digressions that kept us busy all day and turned us
away from studying the initial subject.

Salviati. I remember very well. Returning then to our initial discussion, the cause
of resistance of solids to fracture is certainly within them. Although this resistance
is very strong in the case of push or of direct traction, it is smaller in the case of
a force applied obliquely. So we see, for example, a steel or glass rod resisting to
a longitudinal traction of a thousand pounds, but breaking if fixed to a wall on an
extreme and if a weight of only fifty pounds is applied to the other extreme. We want
to talk about this second resistance, discovering how it varies in prisms and cylinders
of similar or different proportions, but of equal material. In this discussion, I take as
known the principle which governs the behavior of the levers, i.e., that the strength
is to the resistance as the inverse of the ratio of their distances from the fulcrum.

Simplicio. This was first demonstrated by Aristotle in his *Mechanical Problems.*[16]

Salv. Yes, Aristotle had priority, but it seems to me that the quality of the demonstration by Archimedes[17] is far superior because from a single proposition in his
treatise on equilibrium he can derive not only the law of the lever, but also the laws
governing the functioning of most mechanical instruments.

Sagr. But if this principle is the basis of all you want to show us, why don't you
start explaining it to us, if it's not too long?

Salv. Yes, but I prefer to follow a slightly different path from that of Archimedes.
Assuming that equal weights w placed on a two-armed scale with arms of equal
length are in equilibrium, a principle assumed in the same way by Archimedes,[18] I
will show you that not only it is true that different weights w_a e w_b are in equilibrium

© The Author(s), under exclusive license to Springer Nature Switzerland AG 2021   73
A. De Angelis, *Galileo Galilei's "Two New Sciences"*, History of Physics,
https://doi.org/10.1007/978-3-030-71952-4_2

on a balance with arms of different lengths according to the inverse proportion of the weights suspended, but that placing equal weights at equal distances is the same as placing different weights at distances that have an inverse proportion between them compared to the weights. The equilibrium condition is

$$w_a \, a = w_b \, b \, ,\tag{1}$$

which means that the effectiveness of weights (and resistances) is amplified by the distance from the fulcrum of the lever; in short, the effective quantity is the product (moment)

$$M = w \, b \, .\tag{2}$$

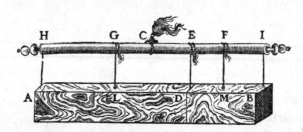

To demonstrate what I say, I draw a prism or a solid cylinder AB, supported at the extremes by two threads, HA and IB. It is evident that if I suspend it by the thread C, placed in the middle of HI, it will remain in equilibrium, since one half of its weight is on one side and the other half on the opposite side of the point of suspension. Now

153   we divide the prism into two unequal parts through the line D. We will have a larger part DA, and a smaller part DB. The system will still remain in equilibrium if we add a thread ED, tied at point E, to support the parts AD and DB. It will still remain in equilibrium if the part of the prism now suspended at both ends with the AH and DE threads will be hooked using a single thread GL, placed in the middle; in the same way, the other part DB does not change its state remaining supported by the FM thread placed in the middle. Removing the threads HA, ED, IB, and leaving the

154   two threads GL and FM, the system will stay in equilibrium. Considering the two weights AD and DB, hanging respectively from the points G and F of the balance GF, it remains only to prove that

$$\frac{w_{DB}}{w_{AD}} = \frac{|GC|}{|CF|} \, ,$$

where w_{DB} e w_{AD} are respectively the weights of the DB and AD parts. Let us normalize the length |AB| to one, and call x the ratio between |AD| and |AB|. We shall have

$$\frac{w_{DB}}{w_{AD}} = \frac{|DB|}{|AD|} = \frac{1-x}{x}.$$

But since $|AF| = 1 - (1-x)/2 = (1+x)/2$, we have

$$\frac{|GC|}{|CF|} = \frac{1/2 - x/2}{(1+x)/2 - 1/2} = \frac{1-x}{x},$$

what we wanted to demonstrate.

155

Having therefore established the principle (1), before proceeding we must consider that these forces, these momenta, these geometric figures, etc., can be considered in abstract, therefore without taking into account the material the bodies are made of, or in concrete, and therefore taking into account the material. In this second approximation, we add matter and therefore weight to the geometric figures.

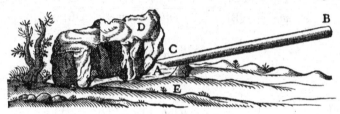

For example, if we take a lever BA, placed on the support E and used to lift a heavy stone D, it is clear, based on the principle just demonstrated, that the strength applied to the end B will be sufficient to balance the resistance of the body D provided that the force applied to the body D is equal to the weight of D times the ratio between the distance AC and the distance CB; this is true if we do not consider other momenta except those of the force B and of the resistance in D, as if the lever were intangible and weightless. If we take into account the weight of the lever, which can be made of wood or iron, the result will be modified and it will therefore be necessary to express it otherwise. It is therefore always necessary to keep in mind under which approximations we are reasoning.

Sagr. I must break my promise not to ask you for digressions, but to concentrate on what will come I must be helped in removing a doubt. It looks like you're comparing the strength applied in B with the total weight of the stone D, but part of this, probably the largest, is supported by the horizontal plane. So...

Salv. I understood very well and you don't have to add anything. Please note that I have not spoken of the total weight of the stone, but only of the force it exercises on point A, the end of the lever BA, which is always smaller than the entire weight of the stone, and varies according to the shape of the stone and to how much it is lifted.

Sagr. I'm satisfied, but I have another problem: how do you figure out which part of the total weight is supported by the underlying plan and which part rests instead on the extremity of the lever?

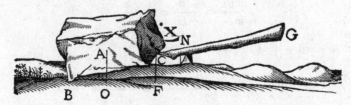

Salv. I can explain it. Let's draw a weight with a center of gravity A; the weight rests with the end B on the horizontal plane and with the other end on the lever CG. Let N be the fulcrum of the lever and G the point where the force is applied. From the center A and from the end C we draw the perpendiculars AO and CF. Calling w the total body weight, w_B the part that weighs on B and w_C what weighs on C,

$$\frac{w_B}{w_C} = \frac{|FO|}{|OB|} \implies \frac{w_B + w_C}{w_C} = \frac{w}{w_C} = \frac{|FB|}{|OB|}$$

and since

$$\frac{w_C}{w_G} = \frac{|GN|}{|NC|}$$

we have

$$\frac{w}{w_G} = \frac{|GN|}{|NC|}\frac{|FB|}{|OB|}.$$

From this, we know also the forces applied in C, in G and in A.

Let us now return to our first purpose. If what we have said so far is clear, it will be also clear that:

Proposition. A prism or a solid cylinder made of glass, iron, wood, or another material, capable of fracture, which can sustain a very large weight when this is applied longitudinally, is sometimes broken by the transverse application of a weight that can be in proportion very small as the length of the cylinder exceeds its thickness.

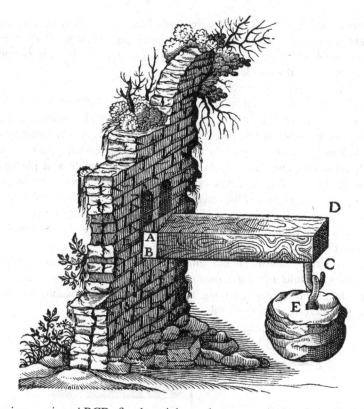

156-
157

Imagine a prism ABCD, fixed at right angle to a vertical wall on its face AB, which holds a weight w_E at the opposite end E. In the point B the wall niche acts as the fulcrum for the lever BC, at the end of which the weight of is applied to the prism; half of the thickness of the solid BA is the other arm of the lever, in which resides the resistance, which consists of the cohesion between the part of the solid that is outside the wall and the part that is inside the wall. From what has been said, it follows that

$$\frac{R}{w_E} = \frac{|CB|}{|AB|/2} = \frac{\ell}{h/2},$$

where R is the resistance, i.e., the force of cohesion between the part of the solid inside the wall and that outside the wall, and the two sides of the section AB are h (the vertical one) and b respectively, so that $\Sigma = ab$ is the section. This is our first proposition. In the case of a cylinder, instead of AB we must consider for h the diameter (and $\Sigma = \pi (h/2)^2$).

It is reasonable to assume that the maximum resistance that the adhesion between the prism and the wall can offer is proportional to the section of the bar:

$$R_{\max} = \sigma_{\lim} \Sigma . \tag{3}$$

In what I said I did not consider the weight of the solid BD. If we want to consider the weight of the prism we must add to the weight w_E half of that of BD. For example, if the latter weighs two pounds and the weight w_E is ten pounds we must consider the weight in C as if it were eleven pounds.

Simp. Why not like it was twelve?

Salv. The weight E hangs from the end of the lever BC with all its weight of 10 pounds. If the weight of BD were hanging at the end it would contribute with all its weight of 2 pounds: but, as you see, the weight of the bar is evenly distributed over the whole length BC, and its parts near the end B weigh less than the most distant ones; adding them up, we have the same situation as if the weight were placed in the center of gravity, which is the center of the BC bar. What matters is the momentum of strength, i.e., the product of the force by the distance from the fulcrum.

158   *Simp.* I understand; and besides, if I am not mistaken, I believe that we would have the same momentum if the double of the weight E were placed in the middle of the lever BC.

Salv. That's right, and we should remember it. We can immediately understand how and in what ratio a rod or a prism of greater breadth than thickness is more resistant when a force is applied along its wider side rather than along the direction in which it is narrower. To clarify this, see the figure.

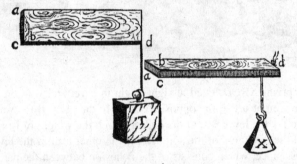

The minimum load that causes the fracture in the figure on the left is the weight w_T of the body T; if you put the axis horizontally, as in the second figure, it will break under the weight w_X, smaller than w_T. This fact becomes evident if we think that the point of application of the resistance is in the first case half of the line ac, and in the second, half of the line cb, and that the distances of the forces inducing the fracture are equal in both cases, and equal to the length $|bd|$; but in the first case the distance of the resistance from the support is greater than in the other case:

$$\frac{w_T}{w_X} = \frac{|ca|/2}{|cb|/2} = \frac{|ca|}{|cb|}.$$

This conclusion is consistent with the idea that the breaking strength is proportional to the number of fibers that must be broken in the material. Therefore we conclude that the same prism resists better to fracture vertically than horizontally, according to the proportion of width to thickness.

Now we want to investigate the resistance of prisms to fracture due to the effect of their own weights. *159-162*

Proposition. For prisms with the same section, the stress exercised on the suspension on a wall is proportional to the square of their lengths.

Calling γ the specific weight of the prism, the resistance R to avoid fracture is given by

$$M = R \frac{h}{2} = \gamma V \frac{\ell}{2} = \gamma bh\ell \frac{\ell}{2} = \gamma bh \frac{\ell^2}{2}.$$

We will now show the proportion according to which the maximum load varies in prisms and cylinders, maintaining the same length and increasing the thickness.

Proposition. In prisms and cylinders of equal lengths, but of different thicknesses (with similar sections), the resistance to fracture grows like the cube of the diameters h of theirs thicknesses, that is their bases.

Indeed[a]

$$M_{\max} = R_{\max} \frac{h}{2} \propto \sigma_{\lim} \Sigma \frac{h}{2} \propto h^3. \tag{4}$$

From what we have shown we can also say that, for the same length,

$$M_{\max} \propto V^{3/2}.$$

In fact, the load due to the weight is proportional to the volume, and the resistance is proportional to the section.

---

[a] Galileo could not take into account the deformation of the material and the consequent proportionality between tensions and deformations later established by Hooke (1635–1703) as a consequence of his famous law. The prism is deformed before it breaks allowing its different sections to rotate around its neutral (i.e., unstressed) axis. Calculations considering this fact were subsequently carried out by Navier (1785–1836) and the results differed from those of Galilei only for numerical factors but not for the dependence on the power of linear dimensions. As a consequence, for example, it is also true in the Navier approximation that, if two nails are fixed to a wall, the one that has double diameter can carry a weight 8 times that of the other (as Salviati will state).

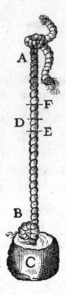

Simp. Before continuing I would like to remove a doubt. You did not take yet into account that resistance in solids decreases as they grow longer, and not only in the transverse sense, but also in the sense of height; as we see, a very long string is e less suitable to hold a great weight than a short one. For this, I believe that a short stick of wood or iron can hold much more weight than a long one.

Salv. I'm sure you're right, if I understand what you mean: a rope, say, forty cubits long, cannot support as much weight as one a cubit or two long.

Simp. That is exactly what I wanted to say. It seems to me very probable.

Salv. Instead, it seems to me improbable, and even wrong. Let's say that on this string AB, hanging at the upper end at point A, the weight C that makes it hung from the other end B breaks. You choose, Simplicio, the point at which the break should occur.

Simp. Let's say that the break occurs at point D.

Salv. Why precisely at D?

Simp. Because the rope wasn't strong enough to hold the weight, let's say a hundred pounds, of the part DB, plus the weight of C.

Salv. So if the rope were stressed by the same hundred pounds of weight in D, it would break at that point.

Simp. It makes sense to me.

Salv. But now tell me: if the weight were attached not to the end of the rope, i.e. in point B, but close to point D, for example in E, and if the rope were fixed not in A, but in a point F near D, would not the point D always feel a weight of one hundred pounds?

Simp. It would, provided you add to the weight C the portion of rope EB.

Salv. Let us therefore suppose that the rope is stretched at point D with a weight of hundred pounds: according to your own admission, it will break. But FE is only a small portion of AB: how can you therefore maintain that the long rope is weaker

than the short one? Give up then this erroneous view that you share with many very
intelligent people, and let's proceed.

*163-
167*

Having shown that with equal length, but with different thickness, the resistances
of prisms and cylinders bound to one side grow according to the proportion of the
cubes of their sides or of the diameters of their bases, and that the effect of a weight
placed at the end grows with the length, that is also the distance from the suspension
point, we conclude that the maximum load at the opposite end to that of suspension
before causing a break is given by:

$$M_{max} \propto \frac{h^3}{\ell} \, . \tag{5}$$

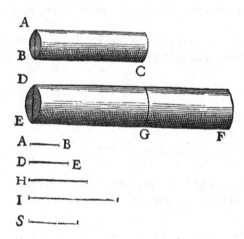

Simp. These statements seem to me not only new, but also unexpected, and very
distant from the opinion I had at first. I would have felt that the resistant momenta
of figures geometrically similar had the same proportion of their resistances.

Sagr. This is the demonstration of the statement that, at the beginning of our
reasoning, seemed obscure to me.

Salv. For a while I used to think, like Simplicio, that the resistances of similar
solids were similar; but a certain casual observation showed me that similar solids
do not exhibit a strength proportional to their size: larger bodies resist less violent
shocks just as tall men are more subject than small children to be injured by a fall.
And falling from same height, we would see a large beam or a column breaking into
pieces, but not a small cylinder of marble.

This observation prompted me to investigate a truly admirable property that I am
going to show you: in solid figures similar to each other the relationships between
resistances and momenta are different.

Simp. You remind me of an argument put by Aristotle[19] in his *Mechanical Problems*, in which he tries to show why a wooden beam becomes weaker and can be
more easily bent as it grows longer, even though the shorter beam is thinner and the

longer one thicker: and, if I remember correctly, he explains it in terms of the simple lever.

Salv. It is true. And since the solution does not seem to eliminate every doubt, Lord Bishop de Guevara, whose commentaries enlightened that work, proposed additional clever speculations to resolve all difficulties, but was still confused about this point: if increasing with the same proportion both the length and the thickness of solid figures these maintain the same resistance to be broken or bent. After a long elaboration I discovered what I am going to expose to you.

Proposition. Among prisms and heavy cylinders geometrically similar, there is only one which under the stress of its own weight lies just on the limit between breaking and not breaking: so that every larger one is unable to carry the load of its own weight and breaks, while smaller ones are able to withstand some additional force tending to break them.

What does it mean that two prisms are similar? It means that the ratios between each pair of homologous dimensions (the sides of the base and the length ℓ) are the same. Let

$$b = \zeta h \; ; \; h = \zeta' \ell,$$

where ζ and ζ' are two proportionality constants. The weight of one of the prisms will be

$$P = \gamma \zeta \zeta'^2 \ell^3,$$

where γ is the specific weight. The momentum M with respect to an end exercised by the weight is equal to the product of the same weight by $\ell/2$:

$$M = \gamma \zeta \zeta'^2 \frac{\ell^4}{2}.$$

On the other hand, the maximum momentum that the prism fixed at one end can tolerate before breaking is, based on equation (4):

$$M_{max} = \frac{\sigma_{lim}}{2} b h^2 = \frac{\sigma_{lim}}{2} \zeta \zeta'^3 \ell^3.$$

By equating the two previous expressions, you get a single value for the maximum length of a prism that does not fall under its own weight (within this set of similar prisms):

$$\ell = \zeta' \frac{\sigma_{lim}}{\gamma}.$$

168-169　　*Sagr.* The proof is clear and concise, and a proposition that at first sight appeared to me unlikely now seems real and unavoidable.

Therefore, to achieve the neutral state between stability and fracture, we will need to alter greatly the ratio between length and thickness of a greater prism by thickening or shortening it, since in two similar prisms the weight is proportional to the third power of any linear dimension, for example ℓ, while its resistance to fracture is proportional to its square.

Salv. Even more, because the question is more difficult; and I know it because I spent some time thinking about it. But now I want to share my results with you.

Problem. Given a cylinder or a prism of length ℓ at the fracture limit, and given a length $\ell' > \ell$, find the minimum diameter of a cylinder (or prism with a similar base) of length ℓ' capable to support its own weight.

Given the previous equation, remembering that $\zeta' = h/\ell$, and using equation (4), we have

$$h' = \frac{\gamma}{\sigma_{\lim}}\ell'^2 = h\sqrt[3]{\frac{\ell'}{\ell}}.$$

From what has already been demonstrated, you can see the impossibility of increasing the size of structures to vast dimensions either in art or in nature, and likewise the impossibility of building ships, palaces, or temples of enormous sizes in such a way that all their parts will hold together. Nor can nature produce trees of extraordinary size because the branches would break down under their own weight; so also it would be impossible to build up the skeletons of men, horses, or other animals so as to hold together and perform their normal functions if these animals were to be increased enormously in height: this increase in height can be accomplished only by employing a material harder and stronger than usual, or by enlarging the size of the bones, thus changing their shape so that the appearance of the animals becomes monstrous. This is perhaps what our wise poet Ariosto had in mind, when he wrote, in describing a huge giant[20]:

> Impossible it is to reckon his height
> So beyond measure is his size.

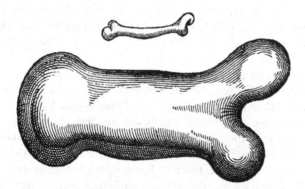

To illustrate this briefly, I have sketched a bone whose natural length has been increased by three times and whose thickness has been multiplied until, for a correspondingly large animal, it would perform the same function which the small bone performs for its small animal. From the figures here shown you can see how out of proportion the enlarged bone appears. Clearly then if one wishes to maintain in a giant the same proportions as in an ordinary man he must either find a harder and

170   stronger material for making the bones, or he must admit a diminution of strength in comparison with men of medium stature; if his height were just increased he would fall and be crushed under his own weight. Whereas, if the size of a body is diminished, the strength of that body is not diminished in the same proportion; indeed the smaller the body the greater its relative strength. Thus a small dog could probably carry on his back two or three dogs of his own size; but I believe that a horse could not carry even one of his own size.

Simp. This may be so; but I am led to doubt it on account of the enormous size reached by certain fish, such as the whales which, I understand, can be ten times as large as an elephant; yet they support themselves.

Salv. Your question, Simplicio, suggests another principle, which had hitherto escaped my attention and which enables giants and other animals of vast size to support themselves and to move about as well as smaller animals do. This result may be secured either by increasing the strength of the bones and other parts intended to carry not only their weight but also the superincumbent load; or, keeping the proportions of the bony structure constant, the skeleton will hold together in the same manner or even more easily, provided one diminishes, in the proper proportion, the weight of the bony material, of the flesh, and of anything else which the skeleton has to carry. It is this second principle that is employed by nature in the structure of fish, making their bones and muscles not merely light but entirely weightless.

Simp. You mean that fish live in water which on account of its density and, as others would say, heaviness diminishes the weight of bodies immersed in it: for this reason, the bodies of fish will be devoid of weight and will be supported without injury to their bones. But this is not all; for although the remainder of the body of the fish may be without weight, there can be no question but that their bones have weight. Take the case of a whale's rib, having the dimensions of a beam; who can deny its great weight or its tendency to go to the bottom when placed in water? One
171   would, therefore, hardly expect these large masses to sustain themselves.

Salv. A very smart objection! And now, in reply, tell me whether you have ever seen fish stand motionless at will under water, neither descending to the bottom nor rising to the top, without the exertion of force by swimming?

Simp. This fact is well known.

Salv. Then the fact that fish are able to remain motionless under water is a conclusive reason for thinking that the material of their bodies has the same specific weight as that of water; accordingly, if in their bodies there are certain parts that are heavier than water there must be others which are lighter, otherwise they would not be in equilibrium. Hence, if the bones are heavier, muscles or other constituents of the body must be lighter in order that their buoyancy may counterbalance the weight of the bones. In aquatic animals therefore circumstances are just reversed from what they are with land animals as, in the latter, the bones sustain not only their own weight but also that of the flesh, while in the former it is the flesh which supports not only its own weight but also that of the bones. We must therefore cease to wonder why these enormously large animals inhabit the water rather than the land, that is to say, the air.

Simp. I am convinced and I only wish to add that what we call land animals ought really to be called air animals, seeing that they live in the air, are surrounded by air, and breathe air.

Sagr. I have enjoyed Simplicio's discussion including both the question raised and its answer. Moreover, I can easily understand that one of these giant fish, if pulled ashore, would not perhaps sustain itself for any great length of time, but would be crushed under its own mass as soon as the connections between the bones gave way.

Salv. I am inclined to your opinion; and, indeed, I almost think that the same thing would happen in the case of a very big ship which floats on the sea without going to pieces under its load of merchandise and armament, but which on dry land and in the air would probably fall apart. But let us proceed.

172

Problem. Given a prism or cylinder of a given weight and length ℓ, and the maximum load w_d that it can carry at an extreme, find a maximum length ℓ' beyond which the cylinder cannot be prolonged without breaking under its own weight.

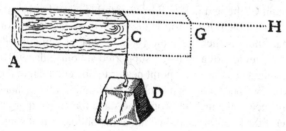

Calling Σ its section, and γ the specific weight,

$$M_{\max} = \gamma \Sigma \frac{\ell^2}{2} + w_d \ell = \gamma \Sigma \frac{\ell'^2}{2},$$

and thus

$$\frac{\ell'^2}{\ell^2} = \frac{w_d \ell + \gamma \Sigma \frac{\ell^2}{2}}{\gamma \Sigma \frac{\ell^2}{2}} = 1 + \frac{2w_d}{\gamma \Sigma \ell}.$$

173

Up to now we have considered the momenta and resistances of prisms and solid cylinders fixed at one end with a weight applied at the other end; three cases were discussed, namely, that in which the applied force was the only one acting, that in which the weight of the prism itself is also taken into consideration, and that in which the weight of the prism alone is taken into consideration. Let's now consider these same prisms and cylinders when supported at both ends or at a single point placed somewhere between the ends.

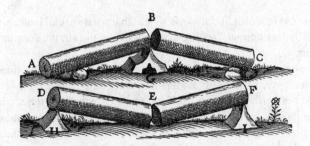

First, I remark that a cylinder carrying only its own weight and having the maximum length beyond which it will break will, when supported either in the middle or at both ends, have twice the length of one which is hanging from a wall and supported only at one end. This is evident because, if we denote the cylinder by ABC and if we assume that one half of it, AB, is the greatest possible length capable of supporting its own weight with one end fixed at B, then, for the same reason, if the cylinder is carried on the point G, the first half will be counterbalanced by the other half BC.

A more difficult problem is the following: neglecting the weight of a solid such as the preceding, find whether the same force or weight which produces fracture when applied at the middle of a cylinder, supported at both ends, will also break the cylinder when applied at some other point nearer to one end than to the other. Thus, for example, if one wished to break a stick by holding it with one hand at each end and applying his knee at the middle, would the same force be required to break it in the same manner if the knee were applied not at the middle, but at some point nearer to one end?

Sagr. This problem, I believe, comes from Aristotle in his *Mechanical Problems.*[21]

174-
177

Salv. His inquiry is not quite the same: he just seeks to discover why a stick may be more easily broken by holding it with one hand at each, far removed from the knee, than if the hands were closer together. Our inquiry calls for something more: we want to know whether, when the hands are retained at the ends of the stick, the same force is required to break it wherever the knee is placed.

Sagr. At first glance, this would appear to be so, because the two lever arms exert, in a certain way, the same momentum, since as one grows shorter the other grows correspondingly longer.

Salv. Now you see how readily one falls into error and what caution and circumspection are required to avoid it. What you have just said appears at first glance highly probable, but on closer examination it proves to be quite far from true; as will be seen from the fact that whether the knee – the fulcrum of the two levers – is placed in the middle or not makes such a difference that, if fracture is to be produced at any other point than the middle, the breaking force at the middle, even when multiplied four, ten, a hundred, or a thousand times, would not suffice.

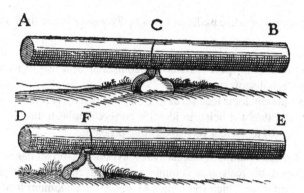

Let AB denote a wooden cylinder which is to be broken in the middle, over the supporting point C, and let DE represent an identical cylinder which is to be broken just over the supporting point F which is not in the middle.

In the first case, considering that the distances AC and CB are identical, also the resistant momenta will be equal:

$$R_B \, |BC| = R_A \, |AC| \implies R_B = R_A \,,$$

where R_A e R_B are respectively the forces applied in A and in B. In the second case we will have from the balance of momenta:

$$R_D \, |DF| = R_E \, |EF| \implies R_D = R_E \, \frac{|EF|}{|DF|} \,.$$

Let us call x the distance DF and $(\ell - x)$ the distance EF. The stick breaks by applying a momentum just above the maximum momentum M_{max} that it can bear:

$$M_{max} = F_D x + F_E (\ell - x) \,.$$

The force in F must balance the sum of the forces in E and D, and thus

$$F_F = \frac{M_{max}}{x} + \frac{M_{max}}{\ell - x} = \frac{M_{max}\ell}{x(\ell - x)} \,. \tag{6}$$

The force will therefore be minimal when $x(\ell - x)$ is maximum, which happens for $x = \ell/2$. Bringing the fulcrum F closer to point D, the sum of forces applied to E and D must be increased to infinity in order to balance or overcome the resistance to F.

Sagr. What should we say, Simplicio? Should we not admit that geometry and mathematics are more powerful than all other tools to refine the spirit and train the mind to think correctly? Plato was not quite right when he wanted his own students first of all to be well prepared in mathematics? I myself knew the properties of leverage and I knew how, by increasing or decreasing its length, strength and endurance increased or decreased, but I was badly wrong in the final answer to the problem.

Simp. I'm actually beginning to understand that the power of logic, despite logic being the guiding tool in speech, cannot be compared with that of mathematics to awaken the spirit to discoveries.[22]

178   *Sagr.* In my opinion, logic teaches us to verify that discoveries lead to conclusions consistently; but I doubt it helps to identify correct demonstrations. Now we can simply apply Eq. (6) to solve a very interesting problem.

Problem. Given the maximum weight that a cylinder or prism can support at its middle point where the resistance is minimum, and given a larger weight, find that point in the cylinder for which this larger weight is the maximum load that can be supported.

Sagr. I understood: and I begin to think that, given that a prism is much more resistant to fracture away from its center, material could be removed from the ends of very large and heavy beams so as to lighten their weight. This fact would be very convenient and useful in the scaffolding of large buildings. It would be interesting to understand which solid figure is equally resistant in all its parts, such that it cannot be broken more easily by a weight in the middle rather than in any other point.

Salv. I was going to give you remarkable information related to this fact. To be more clear, I make a drawing.[23]

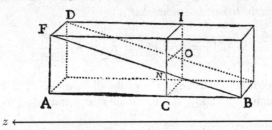

Let DB be a prism. As we have demonstrated,[b] the resistance to fracture in AD by a force applied in the extreme B is smaller than the resistance in CI as the length |CB| is smaller than |BA|. Now let's cut the prism diagonally along the FB line, so that the opposite faces are two triangles, one of which, facing us, is FAB. The new solid thus obtained has different characteristics from those of the prism, and in particular, by applying a force in B, its resistance to fracture in point C is smaller than in point A.

We call z a coordinate from B to A (z therefore varies between 0, in B, and ℓ, in A). Let us consider the maximum resistant momenta in C and in A. The height in C is

$$h_C = \frac{z}{\ell} h ,$$

---

[b] Galileo's pupil Vincenzo Viviani wrote in a note on his copy of the book that in reality this statement does not follow from the previous demonstrations, which however make it plausible. Moreover, Galilei does not explicitly say that this proof neglects the weight of the rod.

and thus

$$M_{C,max} = \frac{\sigma_{lim}bh^2}{2}\left(\frac{z^2}{\ell^2}\right) \; ; \; M_{A,max} = \frac{\sigma_{lim}bh^2}{2}.$$

Hanging a weight w_Q to B, the external momenta in C and A are respectively:

$$M_C = w_Q z \; ; \; M_A = w_Q \ell.$$

Thus

$$\frac{M_{C,max}}{M_{A,max}} = \frac{M_C}{M_A}\left(\frac{z}{\ell}\right).$$

We removed half of the DB prism with a diagonal section, leaving a triangular FBA shape. The initial prism and the new figure show opposite conditions of resistance: while the former is more resistant as you move away from the point of suspension, the latter becomes more fragile. Consequently, it seems reasonable, necessary indeed, that an intermediate cut exists such that, after removing the superfluous, the remaining solid is equally resistant in all its parts. *179-181*

Simp. It is clear that the transition from the major to the minor implies the passage through the equal.

Sagr. Now we must find out how this cut must be made.

Simp. I think it's easy: if sawing the prism diagonally and removing half of the material what remains has opposite characteristics compared to the whole prism, it seems to me that by taking half of that half, i.e., a quarter of the total, the resistance of the remaining figure will be constant in all points, since the gain in a figure is equal to the loss in the other.

Salv. You didn't get it right. You will see that the maximum amount of material that you can remove from the prism without weakening it is not a quarter, but a third.

I will demonstrate that the optimal cut is parabolic. We impose the condition of limit equilibrium

$$M_{max} = w_Q z,$$

and thus

$$\frac{\sigma_{lim}bh^2(z)}{2} = w_Q z;$$

the height $h(z)$ at a generic z is then

$$h(z) = \sqrt{\frac{2w_Q}{\sigma_{lim}b}z},$$

which represents a parabola.[c] And we are removing a third of the weight.

---

[c] Notice that Galilei's figure is inaccurately drawn: the tangent to the curve in B should be vertical.

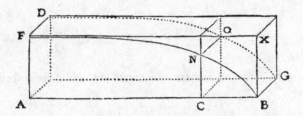

Sagr. The advantages of this weight reduction seem to me so many that it is impossible to list them.

I would like, however, to understand why the weight reduction is exactly one third. I understand very well that the cut along the diagonal removes half the weight. That the parabolic cut takes away a third of the prism, I can believe by trusting Salviati, who always shows to be reliable, but I would prefer to understand scientifically.[d]

Salv. So you want the demonstration that the part of the prism that is removed from this solid that we call parabolic is indeed a third. I know I've already proved it; I will now try to remember the demonstration. First, I need a principle stated by Archimedes in his book *On spirals*[24]:

Lemma. Given an integer number $n > 1$,

$$\frac{1^2 + \ldots + n^2}{n^3} > \frac{1}{3} \tag{7}$$

$$\frac{1^2 + \ldots + (n-1)^2}{n^3} < \frac{1}{3}. \tag{8}$$

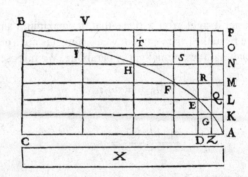

---

[d] We reproduced the demonstration by Galilei instead of using calculus, unknown to him. Calculus-based demonstration is trivial: by calling a the length of the AP segment and b the length of the AC segment, the ABP area is:

$$\int_0^a \frac{b}{a^2} x^2 dx = \frac{ab}{3}.$$

We will prove[e] that, calling S the area of the rectangle ACBP and T the area of the parabolic surface BAP:

$$T = \frac{1}{3}S, \tag{9}$$

and thus the area T' of the curvilinear surface ACB is

$$T' = \frac{2}{3}S = \frac{4}{3}\text{Area(ABP triangle)}.$$

To do it, we shall make the hypothesis that Eq. (9) is not true (and thus $T < S/3$ or $T > S/3$), and we shall demonstrate that this hypothesis implies an absurd.

Let, if possible, be

$$T < \frac{1}{3}S. \tag{10}$$

In this case

$$T + X = \frac{1}{3}S.$$

Divide the BC side into n equal parts as in the figure, so that each of the rectangles that are drawn, for example BO, has an area smaller than X. The area U of the "step-wise" figure circumscribed to BAP (sum of the rectangles : BO + YN + HM + FL + EK + GA) surpasses the area T of the curvilinear part BAP by an area smaller than (BY + YH + HF + FE + EG + GA = BO); but by construction Area(BO) < X, i.e.:

$$U - T < \text{Area(BO)} < X$$

and thus

$$U < T + X = \frac{1}{3}S. \tag{11}$$

It will show now that if it were so, it would also result simultaneously

$$U > \frac{1}{3}S,$$

which is absurd.

In fact, by the properties of parabola, one has

$$|\text{AD}| = k\,|\text{ED}|^2 \;;\; |\text{AZ}| = k\,|\text{GZ}|^2,$$

and thus:

---

[e] Although Galilei writes that he is reproducing a proof due to Archimedes, the first part is different: Archimedes decomposes the surface using triangles, while Galileo uses rectangles.

$$\frac{|ED|^2}{|GZ|^2} = \frac{|AD|}{|AZ|}.$$

Returning to the figure, and calling Area(KE) the area of the rectangle which has opposite vertices K and E, and similarly for other pairs of points, you have (remember also that $|LK| = |AK| = |ZG|$ and that $|AL| = |DE|$):

$$\frac{|DE|^2}{|ZG|^2} = \frac{|AD|}{|AZ|} = \frac{|EL|}{|AZ|} = \frac{|EL|\,|LK|}{|AZ|\,|LK|} = \frac{\text{Area(KE)}}{\text{Area(KZ)}},$$

and also (similarly for further relations):

$$\frac{|AL|^2}{|AK|^2} = \frac{\text{Area(KE)}}{\text{Area(KZ)}}$$
$$\frac{|AN|^2}{|AM|^2} = \frac{\text{Area(MH)}}{\text{Area(LF)}}$$
$$\frac{|AP|^2}{|AO|^2} = \frac{\text{Area(OB)}}{\text{Area(NY)}};$$

and thus

$$\frac{|AL|^2}{\text{Area(KE)}} = \frac{|AK|^2}{\text{Area(KZ)}}$$
$$\cdots$$

(the ratios are the same because it is the same parabola).
    From this, one has

$$\frac{(|AK|^2 + |AL|^2 + |AM|^2 + |AN|^2 + |AO|^2 + |AP|^2)}{\text{Area(KZ + KE + LF + MH + NY + OB)}} = \frac{n|AP|^2}{n\text{Area}(OB)};$$
$$\frac{(|AK|^2 + |AL|^2 + |AM|^2 + |AN|^2 + |AO|^2 + |AP|^2)}{n\text{AP}^2} = \frac{\text{Area(KZ + KE + LF + MH + NY + OB)}}{n\text{Area}(OB)}.$$

Indicating $|AK|$ as d, one has $|AL| = 2d$; $|AM| = 3d$; ...; $|AP| = nd$ and thus the first ratio can be written in a form such that the lemma (7):

$$\frac{(|AK|^2 + |AL|^2 + |AM|^2 + |AN|^2 + |AO|^2 + |AP|^2)}{n\text{AP}^2} > \frac{1}{3}$$

can be applied. From this it follows that the second ratio of the last proportion must also be greater than 1/3, i.e.:

$$\frac{\text{Area(KZ + KE + LF + MH + NY + OB)}}{n\text{Area(OB)}} > \frac{1}{3};$$

but the sum of the rectangles is the circumscribed figure while nArea(OB) is the surface of the ACBP rectangle. We therefore obtained:

$$U > \frac{1}{3}S,$$

in contrast with (11) and thus with (10).

In the same way, an absurd would be obtained assuming BAP $> S/3$, this time considering the inscribed figure constituted by the rectangles VO, TN, ..., QK: a contradiction would be created with the (8). Therefore the area cannot be larger nor smaller than a third of the area of the rectangle BA, and the property (9) can be considered proven; hence the calculation of the area subtended by the parabola.

Sagr. A beautiful, clever demonstration, so much the more so in that it gives us the quadrature of the parabola, proving it to be four-thirds of the inscribed triangle, a fact which Archimedes demonstrated by means of two different, but admirable, series of propositions. This same theorem has also been recently demonstrated by Luca Valerio, the Archimedes of our age; his demonstration can be found in his book dealing with the centers of gravity of solids.[25]

Salv. A book which, indeed, is not to be placed second to any produced by the most eminent geometers of the present and of the past; a book which, as soon as it fell into the hands of our Academician, led him to abandon his own researches along these lines; for he saw how happily everything had been treated and demonstrated by Valerio.[26]

Sagr. When I was informed of this event by the Academician himself, I begged him to show the demonstrations which he had discovered before seeing Valerio's book; but in this I did not succeed. *185*

Salv. I have a copy of them and will show to you: you will enjoy the diversity of the methods employed by these two authors in reaching and proving the same conclusions, although both are equally correct.

Sagr. I shall be much pleased to see them and will consider it a great favor if you will bring them to our regular meeting. But in the meantime, considering the strength of a solid formed from a prism by means of a parabolic section, would it not, in view of the fact that this result promises to be both interesting and useful in many mechanical operations, be a fine thing if you were to give some quick and easy rule by which a mechanic might draw a parabola upon a plane surface?

Salv. There are many ways to draw a parabola, and two are particularly simple.

One is really marvelous, because in less than the time it takes to draw in a precise way with the compass four or six circles of different sizes I can draw thirty and forty parabolic lines not less precise, thin and clean than the circumferences of these circles. I use a perfectly spherical ball of bronze, not bigger than a walnut; this, thrown on a mirror of metal inclined with respect to the horizontal, traces a very thin and very precise parabolic line, wider or narrower depending on the inclination of the mirror. This tells us among other things that the projectile motion has a parabolic trajectory. It was our friend who first observed this phenomenon, bringing back the demonstration in his book on the motion, that we will see together in ours next *186* meeting. Because the ball leaves clear traces of its parabolic motion above the mirror it is best to heat it and moisten it for a long time in the hand.

The other way it works as follows. We fix two nails high on a wall on a horizontal line, distant from each other twice the width of the rectangle on which we want to trace the semiparabola, and from these two nails we make to hang a thin chain, so long that the curve shaped under gravity extends as long as the length of the prism. This chain describes a parabola[f]. Tracing the path from the chain over the wall, we will draw an entire parabola, which we will divide into two equal parts with a plumb line hanging from the middle of the two nails. It is easy then to transfer this line over the opposite faces of the prism, and also a craftsman without great skills will be able to do it.

In addition, it would be possible without effort to trace a parabola on the face of the prism, joining a series of points drawn with the help of the proportional compass invented by our friend.[g]

We have shown so far many conclusions concerning the resistance of solids to fracture, assuming the longitudinal strength is known. On the basis of these we can continue our journey finding different and new conclusions and demonstrations among the infinite that are in nature. I would like now, as a final goal of our reasoning today, to add a consideration on the resistance of solids, that technique, and even more nature, use in a thousand operations in which greater sturdiness is needed without increasing weight. This is the case, for example, of bird bones and tubes, which are light and very resistant to bending or breaking. A straw, which supports one spike heavier than all its stem, if it were massive, even if made of the same quantity of material, would be much less resistant. Artisans know well that an empty rod or a rod of wood or metal is much more solid than if it were, with the same weight and the same length, massive, and therefore also thinner; the practice suggests building lances internally empty because they are strong and light.

187-
189

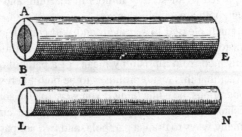

We will show that

---

[f] Indeed the curve thus obtained is a catenary and not a parabola, but it approximates a parabola.

[g] Galilei designed this instrument capable of carrying out complex mathematical and geometric operations and had it built in Padua in 1597 by his worker Marcantonio Mazzoleni, brother of Mario, chair of natural philosophy at the University. The instrument is described in the booklet *The operations of the geometric and military compass*, published in Padua in 1606 and dedicated to Cosimo II Medici. The proportional compass was very successful and Galilei had it mass produced to sell it. Here Galilei has just advertised a commercial product of his.

Proposition. The resistances of two cylinders of equal and equally long sections, one of which is empty and the other massive, stand between them as their diameters.

Calling D the external diameter of the hollow cylinder, and $d < D$ the diameter of a full cylinder of the same weight, the maximum momenta, that we call respectively M and m, are:

$$M = R\frac{D}{2} \; ; \; m = R\frac{d}{2}.$$

Thus, for the same amount of material, one has for the maximum momenta

$$\frac{M}{m} = \frac{D}{d} > 1 \,,$$

quod erat demonstrandum.[h] The robustness of a pipe is, therefore, greater than the strength of a solid cylinder of the same weight and of the same length made of the same material, in proportion to the diameters. The previous relations also give us a way to calculate what happens in a general case, between unequal weight and less deeply excavated tubes.[27]

The second day ends.[i]

---

[h] As explained in footnote (a) on this day, Galilei's calculation neglects the elasticity of the material. In this case, unlike the one discussed in the previous note, the more accurate calculation made by Navier does not have the same simplicity and the same dimensional dependence as that of Galilei.

[i] Galilei probably wanted to add some material at this point. The traditional "entr'acte" at the end of the day is missing and the beginning of the next day is abrupt, with the sudden appearance of the treatise in Latin written by the Academician (which had only been briefly discussed).

Day Three

(Other New Science, on Local Motion)

[Salviati, reading from the Latin treatise on local motion by the Author.] I will develop a completely new science concerning a very old topic. There is perhaps no older argument of the movement among those faced by the philosophers; yet it seems to me that many essential aspects do not have not even been noticed, much less demonstrated. Some simple aspects have been observed, such as the fact that heavy bodies fall and accelerate; but the mathematical law that describes such an acceleration has not yet been found. Nobody indeed, as far as I know, has demonstrated yet that spaces covered in equal times by a body that falls starting from rest are between them as the odd numbers following unity. It has been observed that projectiles trace a curved line, but no one has shown yet that this line is a parabola. We shall demonstrate these and other things, not less worthy of being known; more importantly, all this will open the way to one new and great science of which these demonstrations will be the elements, a science thanks to which minds sharper than mine will be able to penetrate even deeper recesses.[a]*

190-196

This discussion is divided into three parts: the first deals with uniform motion; the second with naturally accelerated motion; the third concerns the violent movement of bullets.

UNIFORM MOTION

To treat constant or uniform motion we need a single definition.

Definition. [28] *By steady or uniform motion, I mean one in which the distances traversed by a moving particle during any equal intervals of time, are themselves equal.*

---

[a] We will soon see that this relationship means that the space traveled by a body starting from rest is proportional to the square of the elapsed time. Galilei had already discovered this relationship in Padua in 1604 (as one can read in his notes), and used it in the *Dialogue*. Failure to quote the *Dialoge* is probably due to the fact that that work had been banned by the Church.

© The Author(s), under exclusive license to Springer Nature Switzerland AG 2021
A. De Angelis, *Galileo Galilei's "Two New Sciences"*, History of Physics,
https://doi.org/10.1007/978-3-030-71952-4_3

From this definition it follows that:

$$s(t) = v_0 t \,, \tag{19}$$

with v_0 the constant speed.

Salv. We have seen so far what our author wrote about uniform motion. Let us now turn to a more difficult problem: naturally accelerated motion, like the one generally experienced by falling bodies. Here are the title and the introduction.

197-
198

NATURALLY (UNIFORMLY) ACCELERATED MOTION

We have just discussed the properties of uniform motion; the accelerated motion remains to be considered. First of all, it seems desirable to find a definition that fits the natural phenomena. Indeed anyone can invent an arbitrary type of movement and discuss its properties; so, for example, some have imagined spirals and other complicated curves called concoids not described by motions found in nature, and established the properties of such curves. We decided instead to consider the phenomenon of the accelerated fall of bodies like it occurs in nature, and give a definition of accelerated motion reproducing the essential characteristics of the accelerated motions observed. And in the end, after repeated efforts, we believe we succeeded: we have confidence that the experimental results are in agreement with the properties we will demonstrate. Finally, in the investigation of naturally accelerated motion[b] we followed the habit of nature itself, in all its various aspects, assuming that this is the simplest. I think everybody agrees that fish and birds swim and fly instinctively in the easiest possible way. So, when I look at a stone initially at rest, that falls from an elevated position and continuously acquires new increments of speed, why should I doubt that such increases occur in the simplest possible way? No increase is simpler than what is always repeated in the same way. Just like the uniformity of movement is defined by moving through equal spaces in equal times (so we call a movement uniform when equal distances are crossed during equal time intervals), we can imagine a uniform and continuously accelerated movement when, during any time interval, speed has equal increments. Then, dividing time into equal intervals, counting it from the beginning of the descent, the amount of speed acquired during the first two time intervals will be twice that acquired during the first interval; the amount added during three intervals of time will be triple, and so on. To be clearer, if a body should continue its movement with the same speed that it had acquired during the first interval of time and should maintain the same uniform speed, then its movement would be twice as slow than if its speed had been acquired during two intervals of time. And so, apparently, we are not wrong if we say that:

$$v(t) = at \,, \tag{20}$$

---

[b] Galilei uses the expression "motus naturaliter acceleratus" also in the title of the section of the treatise, meaning by this that his main objective is the description of free fall. In our translation we have indifferently used the expressions "naturally accelerated" and "uniformly accelerated".

with a being a constant. The definition of motion we are about to discuss can be formulated as follows: a particle is uniformly accelerated when, starting from rest, it acquires equal increments of speed in equal time intervals.

Sagredo. Although I can't offer any rational objection to this or to any other definition, since all definitions are arbitrary, I may nevertheless be allowed to doubt whether such a definition as the above, established abstractly, corresponds to and describes the kind of accelerated motion that we meet in nature in the case of freely falling bodies. And since the Author apparently maintains that the motion described in his definition is that of freely falling bodies, I would like to clear my mind of certain difficulties so that I may later apply myself more attentively to the propositions and their demonstrations.

Salv. It is good that you and Simplicio raise these difficulties. They are, I imagine, the same which occurred to me when I first saw this treatise, and which were removed either by discussing with the Author himself, or by turning the matter over in my own mind.

Sagr. When I think of a heavy body falling from rest, I see it starting with zero speed and gaining speed in proportion to the time from the beginning of the motion. Under such a motion it would, for instance, in eight pulse beats acquire eight degrees of speed; at the end of the fourth beat having acquired four degrees; at the end of the second, two; at the end of the first, one. Since time is divisible without limit, if the earlier speed of a body is less than its present speed by a constant ratio, then there is no degree of speed however small (or, one may say, no degree of slowness however great) with which we may not find this body traveling after starting from infinite slowness, i.e., from rest. So that if that speed which it had at the end of the fourth beat was such that, if kept uniform, the body would traverse two miles in an hour, and if keeping the speed which it had at the end of the second beat, it would traverse one mile an hour, we must infer that, as the instant of departure is more and more nearly approached, the body moves so slowly that, if it kept on moving at this rate, it would not traverse a mile in an hour, or in a day, or in a year or in a thousand years; indeed, it would not traverse a span in an even greater time; a phenomenon which baffles the imagination, while our senses show us that a heavy falling body suddenly acquires great speed.

Salv. This is one of the difficulties that I also experienced at the beginning, but that I shortly afterward removed; and I could do it thanks to the very experiment which creates your difficulty. You say that the experiment appears to show that immediately after a heavy body starts from rest it acquires a very considerable speed: and I say that the same experiment makes clear the fact that the initial motion of a falling body, no matter how heavy, is very slow and gentle. Place a heavy body upon a yielding material, and leave it there without any pressure except that owing to its own weight. It is clear that if one lifts this body by a cubit or two and lets it fall upon the same material, it will, with this impulse, exert a new and greater pressure than that caused by its mere weight. This effect is brought about by the weight of the falling body together with the velocity acquired during the fall, an effect which will be greater and greater according to the height of the fall, i.e., depending on the increase of the velocity of the falling body. From the intensity of the blow we can thus accurately

199

estimate the speed of a falling body. But tell me: is it not true that if a block be allowed to fall upon a stake from a height of four cubits and drives it into the ground, say, four inches, that coming from a height of two cubits will drive the stake a much less distance, and from the height of one cubit a still less distance; and finally if the block is lifted only one inch how much more will it accomplish than if merely laid on top of the stake without percussion? Certainly very little. If it is lifted only the thickness of a leaf, the effect will be altogether imperceptible. And since the effect of the blow depends upon the velocity of this striking body, can anyone doubt that the motion is very slow and the speed more than small whenever the effect of the blow is imperceptible? See now the power of truth: the same experiment which at first glance seemed to show one thing, when more carefully examined, assures us of the contrary. But without depending upon the above experiment, which is certainly very conclusive, it seems to me that it ought not to be difficult to establish such a fact just by reasoning. Imagine a heavy stone held in the air at rest; the support is removed and the stone set free. Since the stone is denser than air it begins to fall, and not with uniform motion but slowly at the beginning and with a continuously accelerated motion. Now since velocity can be increased and diminished without limit, what reason is there to believe that such a moving body starting with infinite slowness, that is, from rest, immediately acquires a speed of ten degrees rather than one of four, or of two, or of one, or of a half, or of a hundredth; or, indeed, of any of the infinite number of small values of speed? I hardly think you will refuse to grant that the gain of speed of the stone falling from rest follows the same sequence as the diminution and loss of this same speed when, by some impelling force, the stone is thrown to its former elevation: but even if you do not grant this, I do not see how you can doubt that the ascending stone, diminishing in speed, must pass through every possible degree of slowness before coming to rest.

Simplicio. But if the number of degrees of greater and greater slowness is limitless, they will never be all exhausted, therefore such an ascending heavy body will never reach rest, but will continue to move without limit at an always slower rate; and this is not what is observed.

Salv. This would happen if the moveable were to hold itself for any time in each degree; but it merely passes there, without remaining beyond an instant. And since each time interval however small may be divided into an infinite number of instants, these will always be sufficient in number to correspond to the infinite degrees of diminished velocity. That such a heavy rising body does not remain for any length of time at any given degree of velocity is evident from the following: if, some time interval having been assigned, the body moves with the same speed in the last as in the first instant of that time interval, it could from this second degree of elevation be in a like manner raised through an equal height, just as it was transferred from the first elevation to the second, and by the same reasoning would pass from the second to the third and would finally continue in uniform motion forever.

Sagr. From these considerations, it appears to me that we may obtain a proper solution to the problem discussed by philosophers, namely, what causes the acceleration in the natural motion of heavy bodies. Since, as it seems to me, the motion impressed by the agent projecting the body upwards diminishes continuously, this force, so long as it was greater than the contrary force of gravitation, impelled the body upwards; when the two are in equilibrium the body ceases to rise and passes through the state of rest in which the impressed impetus is not destroyed, but only its excess over the weight of the body has been consumed – the excess which caused the body to rise. Then as the diminution of the outside impetus continues, and gravity gains the upper hand, the fall begins, but slowly at first on account of the opposing impetus, a large portion of which still remains in the body; but as this continues to diminish it also continues to be more and more overcome by gravity, hence the continuous acceleration of motion.

Simp. The idea is clever, yet more subtle than sound; for even if the argument were conclusive, it would explain only the case in which a natural motion is preceded by a violent motion, in which a portion of the external force still remains active; but where there is no such remaining portion and the body starts from a state of rest, the cogency of the whole argument fails.

Sagr. I believe that you are mistaken, and that your distinction between cases is superfluous. But, tell me, can't a projectile receive from the projector either a large or a small force such as will throw it to a height of a hundred cubits, and even twenty or four or one?

Simp. Certainly yes.

202

Sagr. Therefore this impressed force may exceed the resistance of gravity so slightly as to raise it only an inch; and finally the force of the projector may be just large enough to exactly balance the resistance of gravity so that the body is not lifted at all but merely sustained. When one holds a stone in his hand does he do anything but give it a force impelling it upwards equal to the force of gravity drawing it downwards? And do you not continuously impress this force upon the stone as long as you hold it in the hand? Does it perhaps diminish with the time during which one holds the stone? And what does it matter whether this support which prevents the stone from falling is provided by one's hand or by a table or by a rope from which it hangs? Certainly nothing at all. You must conclude, therefore, that it makes no difference whatever whether the fall of the stone is preceded by a period of rest which is long, short, or instantaneous, provided only the fall does not take place so long as the stone is balanced by a force opposed to its weight and sufficient to hold it.

Salv. This does not seem to be the proper time to investigate the cause of the acceleration of natural motion, Various opinions have been expressed by various philosophers, some explaining it by attraction to the center, others by repulsion between the very small parts of the body, while still others by a stress in the surrounding medium which closes in behind the falling body and drives it from one of its positions to another. All these fantasies, and others too, ought to be examined; but it is not really worthwhile to do it now. At present, the purpose of our Author is merely to investigate and to demonstrate some of the properties of accelerated motion (whatever the cause of this acceleration may be) – meaning thereby a motion, such that its velocity goes on increasing after departure from rest, in simple proportionality to time, which is the same as saying that in equal time intervals the body receives equal increments of velocity; and if we find the properties of accelerated motion which will be demonstrated later are realized in freely falling and accelerated bodies, we may conclude that the assumed definition includes such a motion of falling bodies and that their speed goes on increasing as the time and the duration of the motion.

203

Sagr. As far as I see at present, the definition might have been formulated a little more clearly perhaps without changing the fundamental idea: namely, uniformly accelerated motion is such that its speed increases in proportion to the space traversed; so that, for example, the speed acquired by a body in falling four cubits would be double that acquired in falling two cubits and this latter speed would be double that acquired in the first cubit. Because there is no doubt but that a heavy body falling from the height of six cubits has, and strikes with, a speed double that it had at the end of three cubits, and triple that which it had at the end of one.

Salv. It is very comforting to me to have had such a companion in error; moreover, let me tell you that your proposition seems so highly probable that our Author himself admitted, when I advanced this opinion to him, that he had for some time shared the same fallacy. But what most surprised me was to see that two propositions so inherently probable that they commanded the consensus of most could be proven impossible in a few simple words.

Simp. I am one of those who accept the proposition, and believe that a falling body acquires impetus in its descent, its velocity increasing in proportion to space, and that the momentum of the falling body is doubled when it falls from a double height; these propositions, it appears to me, ought to be conceded without hesitation or controversy.[c]

Salv. And yet it is false and impossible since in that case the motion should be completed instantaneously; and here is a very clear demonstration of it. If the velocities are in proportion to the distance traversed, or to be traversed, then these distances are traversed in equal intervals of time; if, therefore, the velocity with which the falling body traverses a space of eight feet were double that with which it covered the first four feet (just as the one distance is double the other), then the time intervals required for these passages would be equal. But for the same body to fall eight feet and four feet in the same time is possible only in the case of instantaneous

204

motion. Observation shows instead that the motion of a falling body takes time, and less of it in covering a distance of four feet than of eight feet; therefore it is not true that its velocity increases in proportion to the space. The falsity of the other proposition may be shown with equal clearness. If we consider a single striking body the difference of momentum in its blows can depend only upon difference of velocity; if the striking body falling from a double height were to deliver a blow of double momentum, it would be necessary for this body to strike with a doubled velocity; but with this doubled speed it would traverse a doubled space in the same time interval. Observation however shows that the time required for fall from the greater height is longer.

Sagr. You present these recondite matters with too much evidence and ease; this great facility makes them less appreciated than they would be had they been presented

---

[c] Here Simplicio is not an Aristotelic.

in a more abstruse manner. In my opinion, people esteem less that knowledge which they acquire with little labor than that acquired through long and obscure discussion.

Salv. If those who demonstrate with brevity and clearness the fallacy of many popular beliefs were treated with contempt instead of gratitude the injury would be quite bearable; but on the other hand, it is very unpleasant and annoying to see people, who claim to be peers of anyone in a certain field of study, take for granted certain conclusions which later are quickly and easily shown by another to be false. I do not describe such a feeling as a kind of envy, which usually degenerates into hatred and anger against those who discover such fallacies; I would call it a strong desire to maintain old errors, rather than accepting newly discovered truths. This desire sometimes induces some people to unite against truths, although at heart believing in them, merely to lower the esteem in which certain others are held by the unthinking crowd. Indeed, I have heard from our Academician many such fallacies held as true but easily refutable; some of these I have in mind.

Sagr. You must not withhold them from us, but, at the proper time, tell us about them even though an extra session will be necessary. But now, continuing the thread of our talk, it would seem that up to now we have established the definition of   *205* uniformly accelerated motion which is expressed as follows:

Definition. *A motion is said to be naturally or uniformly accelerated when, starting from rest, the speed receives equal increments in equal times:*

$$v(t) = at \, , \tag{21}$$

with a a constant (called acceleration).

Salv. This definition established, the Author makes a single assumption, namely:

Postulate. *The speeds acquired by a body moving down along planes of different inclinations are equal when the heights of these planes are equal.*

By the height of an inclined plane, we mean the length of the perpendicular from the upper end of the plane to the horizontal line drawn through the lower end of the same plane. Thus, to illustrate, let the line AB be horizontal, and let the planes CA and CD be inclined to it; then the Author calls the perpendicular CB the "height" of the planes CA and CD; he supposes that the speeds acquired by a body descending along the planes CA and CD to the terminal points A and D are equal since the heights of these planes are the same, CB; and this speed is that which would be acquired by the body falling from C to B.

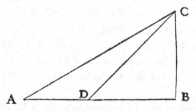

Sagr. Your assumption appears to me so reasonable that it ought to be conceded without question, provided of course there are no outside resistances, and that the planes are smooth, and that the moving body is perfectly round. All resistance and opposition having been removed, my reason tells me at once that a heavy and perfectly round ball descending along the lines CA, CD, CB would reach the terminal points A, D, B, with equal speed.

Salv. Your words are very plausible; but I hope by experiment to increase the plausibility to an extent that shall be little short of a demonstration.

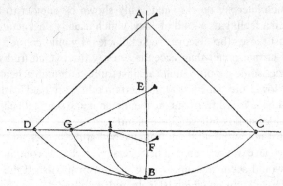

206-
214

Imagine that this page represents a vertical wall, with a nail driven into it, and from the nail let there be suspended a lead bullet of one or two ounces using a fine vertical thread, AB, say from four to six feet long. Draw on this wall a horizontal line DC, at right angles to the vertical thread AB, which hangs about two inches in front of the wall. Now bring the thread AB with the attached ball into the position AC and set it free; it will be observed to descend along the arc CBD, to pass point B, and to travel along the arc BD, till it almost reaches the horizontal CD, a slight shortage being caused by the resistance of the air and the string; from this we may rightly infer that the ball in its descent through the arc CB acquired an impetus on reaching B, that was just sufficient to carry it through a similar arc BD to the same height. Having repeated this experiment many times, let us now drive a nail into the wall close to the perpendicular AB, say at E or F, so that it projects out some five or six inches so that the thread, carrying the bullet through the arc CB, may strike upon the nail E when the bullet reaches B, and thus compel it to traverse the arc BG, described about E as the center. From this we can see what can be done by the same impetus that previously starting at the same point B carried the same body through the arc BD to the horizontal CD. Now you will observe that the ball swings to the point G in the horizontal, and you would see the same thing happen if the obstacle were placed at some lower point, say at F, about which the ball would describe the arc BI, the rise of the ball always terminating exactly on the line CD. But when the nail is placed so low that the remainder of the thread below it will not reach to the height CD (which would happen if the nail were placed nearer B than to the intersection of AB with the horizontal CD) then the thread leaps over the nail and twists itself about it.

This experiment leaves no room for doubt as to the truth of our supposition; since the two arcs CB and DB are equal and similarly placed, the impetus acquired by the fall through the arc CB is the same as that gained by fall through the arc DB, as by the fall through the arcs GB and IB.

Sagr. This argument seems to me so conclusive and the experiment so well adapted to establish the hypothesis that we may, indeed, consider it as demonstrated.

Salv. I don't want us to make any further assumptions about this, since we will apply this principle mainly to motions that occur on flat surfaces, and not on curved surfaces, along which the acceleration varies differently from what we have for inclined planes. For now, consider this as a postulate, the truth of which will be established when we discover that what we can infer from it corresponds and agrees perfectly with our observations. Having assumed this single principle, the Author passes next to propositions that he demonstrates; the first of these is as follows:

Proposition (theorem). *The time in which any space is traversed by a body starting from rest and uniformly accelerated is equal to the time in which that same space would be traversed by the same body moving at a uniform speed whose value is half of the final speed.*

Since $v(t) = at$, the time-averaged speed in the uniformly accelerated motion of a body starting from rest and ending its motion at a final speed v_f, will be equal to $v_f/2$. The movement can be thought of as a sequence of uniform movements with velocity $v(t)$, and therefore

$$s(t) = \frac{v_f}{2}t. \tag{22}$$

Proposition (theorem). *The spaces described by a body that falls starting from rest and moves with a uniformly accelerated motion stand between them like the squares of the time intervals employed in traveling these distances.*

$$s(t) = \frac{v_f}{2}t$$

from which, considering that at time $t/2$ the speed is half of the final speed,

$$\frac{v_f}{2} = v\left(\frac{t}{2}\right) = a\frac{t}{2},$$

and finally we have

$$s(t) = \frac{1}{2}at^2. \tag{23}$$

Corollary. *If we consider time intervals of equal length from the beginning of the movement, the spaces traveled in successive intervals will stay between them in the same relationship as the succession of odd numbers: 1, 3, 5, ...*

In fact, expressing the odd numbers starting from the unit as $(2j - 1)$, with $j = 1, 2, ...$ (the sequence of natural numbers), we have

$$1 = 1^2$$
$$1 + 3 = 2^2$$
$$1 + 3 + 5 = 3^2$$
$$\cdots$$
$$1 + 3 + \ldots + (2n - 1) = n^2.$$

If, therefore, during equal time intervals the velocity increases linearly (and therefore proportionally to the sequence of natural numbers), the increases in the distances traveled during these equal time intervals stand between them as the odd numbers starting with the unit, and the total space traveled is proportional to the square of time, and vice versa.

Simp. I am convinced that matters are as described, once having accepted the definition of uniformly accelerated motion. But as to whether this acceleration is that which one meets in nature in the case of falling bodies, I am still doubtful; and it seems to me, not only for my own sake but also for all those who think as I do, that this would be the proper moment to introduce one of those experiments—and there are many of them, I understand—which illustrate in several ways the conclusions reached.

Salv. The request that you, as a man of science, make, is very reasonable; for this is the custom—and properly so— in those sciences where mathematical demonstrations are applied to natural phenomena, as is seen in the case of perspective, astronomy, mechanics, music, and others where the principles, once established by well chosen experiments, become the foundations of the entire superstructure. I hope therefore it will not appear to be a waste of time if we discuss at considerable length this first and most fundamental question upon which hinge numerous consequences of which we have in this book only a small number, placed there by the Author, who has done so much to open a pathway hitherto closed to speculative minds. So far as experiments go they have not been neglected by the Author; and often, in his company, I have attempted in the following manner to assure myself that the acceleration actually experienced by falling bodies is that above described.

On the edge of a piece of wooden molding or scantling, about 12 cubits long, half a cubit wide, and three inches thick, a channel was cut a little more than one inch, very straight, smooth, and polished. Having lined it with parchment, also as smooth and polished as possible, we rolled along it a hard, smooth, and very round bronze ball. Having placed this aboard in a sloping position, by lifting one end some one or two cubits above the other, we rolled the ball, as I was just saying, along the channel, noting, in a manner presently to be described, the time required to make the descent. We repeated this experiment more than once in order to measure the time with an accuracy such that the deviation between two observations never exceeded one tenth of a pulse-beat.

Having performed this operation and having assured ourselves of its reliability, we now rolled the ball only one quarter the length of the channel; and having measured the time of its descent, we found it precisely one half of the former. Next, we tried other distances, comparing the time for the whole length with that for the half, or with that for two thirds, or three fourths, or indeed for any fraction; in such experiments, repeated a full hundred times, we always found that the spaces traversed were to each other as the squares of the times, and this was true for all inclinations of the plane, i.e., of the channel, along which we rolled the ball. We also observed that the times of descent, for various inclinations of the plane, bore to one another precisely that ratio that, as we shall see later, the Author had predicted and demonstrated for them.

For the measurement of time, we employed a large vessel of water placed in an elevated position; to the bottom of this vessel was soldered a pipe of small diameter releasing a thin jet of water, that we collected in a small glass during the time of each descent, whether for the whole length of the channel or for a part of its length; the water thus collected was weighed, after each descent, on a very accurate balance; the differences and ratios of these weights gave us the differences and ratios of the times, and this with such accuracy that although the operation was repeated many, many times, there was no appreciable discrepancy in the results.

Simp. I would like to have been present at these experiments; but feeling confidence in the care with which you performed them, and in the fidelity with which you relate them, I am satisfied and accept them as true and valid.

Salv. Then we can proceed without discussion.

Corollary. *Starting from any initial point, if we take any two distances, traversed in arbiitrary time intervals, these time intervals bear to one another the same ratio as one of the distances to the mean proportional of the two distances.*
    From

$$s_1 = at_1^2$$

and

$$s_2 = at_2^2$$

we have, by dividing the corresponding sides of the equation,

$$\frac{t_1}{t_2} = \sqrt{\frac{s_1}{s_2}} = \frac{\sqrt{s_1 s_2}}{s_2}$$

($\sqrt{s_1 s_2}$ is the geometric mean, or mean proportional, between s_1 and s_2).

Comment. The aforementioned corollary also applies to inclined planes at any angle, since we assume that along these planes the speed increases in the same ratio, i.e. in proportion to time or, if you prefer, as the sequence of the natural numbers.

Salv. Here, Sagredo, I should like, if it be not too tedious to Simplicio, to interrupt for a moment the present discussion to make some additions based on what has already been proved and of what mechanical principles we have already learned from our Academician. This addition I make for the better establishment on logical and experimental grounds, of the principle that we have above considered; and what is more important, for the purpose of deriving it geometrically, after first demonstrating a single lemma that is fundamental in the science of motion.

215-
223

Sagr. If the advance that you propose to make is such as will confirm and fully establish these sciences of motion, I will gladly devote to it any length of time. Please, satisfy the curiosity that you have awakened in me concerning your proposition; and I think that Simplicio agrees.

Simp. Of course.

Salv. Let us first consider this notable fact that the speeds of a body moving through the same length vary with the inclination of the plane. The speed reaches a maximum along a vertical direction, and for other directions diminishes as the plane diverges from the vertical.

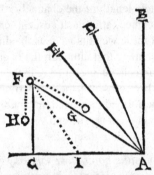

For the sake of greater clearness erect the line AB perpendicular to the horizontal AC; next draw AD, AE, AF, etc., at different inclinations to the horizontal. Then I say that all the velocity of the falling body is along the vertical and is a maximum when it falls in that direction; the velocity is smaller along DA and still less along EA, and even less yet along the more inclined plane FA. Finally, on the horizontal plane the velocity vanishes. The body finds itself in a condition of indifference as to motion or rest; has no inherent tendency to move in any direction, and offers no resistance to being set in motion. For just as a heavy body or system of bodies cannot of itself move upwards, or recede from the common center toward which all heavy things tend, so it is impossible to assume any motion other than one which carries it nearer to the aforesaid common center. Hence, along the horizontal, by which we understand a surface, every point of which is equidistant from this same common center, the body will acquire no velocity whatever.

I need to explain something that our Academician wrote when in Padua, embodying it in a treatise on mechanics prepared solely for the use of his students, and proving it conclusively when considering the origin and nature of that marvelous machine, the screw. He demonstrated how the change of speed along planes of different inclinations is different. Given the inclined plane AF, for example, and taking as its elevation above the horizontal the line FC, along which the impetus of a heavy body and its momentum in descent is maximum, we want to know the ratio that this impetus has to the impetus of the same moveable along the inclined line FA. I say that this ratio is inverse to that of the said lengths.

It is clear that the force acting on a body in descent is equal to the minimum force sufficient to hold it at rest. To measure this force, I propose to use the weight of another body. Let us place upon the plane FA a body G connected to the weight H by means of a cord passing over the point F; then the body H will ascend or descend, along the perpendicular, the same distance which the body G ascends or descends along the inclined plane FA. This is clear. If we consider the motion of the body G from A to F, in the triangle AFC to be made up of a horizontal component AC and a vertical component CF, and remember that this body experiences no resistance to motion along the horizontal (because by such a motion the body neither gains nor loses distance from the common center of the heavy things), it follows that resistance is met only in consequence of the body rising through the vertical distance $|CF|$. Since then the body G when moving from A to F offers resistance only in so far as it rises through the vertical distance $|CF|$, while the other body H must fall vertically through the entire distance $|FA|$, and since this ratio is maintained whether the motion is large or small, the two bodies being connected at a fixed mutual distance, we are able to assert that, in case of equilibrium the velocities in their tendency to motion, i.e., the spaces that would be traversed by them in equal times if they would move, must be in the inverse ratio to their effective weights. This is what has been demonstrated in every case of mechanical motion. In conclusion, in order to hold the weight in G, w_G, at rest, one must give H a weight w_H smaller in the same ratio as the distance $|FC|$ is smaller than $|FA|$. If

$$|FA| : |FC| = w_G : w_H$$

then equilibrium will occur, that is, the weights H and G will have the same impelling forces, and the two bodies will come to rest. And since we agree that impetus, energy, momentum, tendency to motion of a body[d] is as great as the force or least sufficient to stop it, and we have found that the weight in H is capable of preventing motion in the weight in G, it follows that the smaller weight w_H whose entire force is along the perpendicular, FC, will be an exact measure of the component of force that the larger weight w_G exerts along the plane FA. But the measure of the total force on the body G is its own weight, since to prevent its fall it is only necessary to balance it with an equal weight, provided this second weight is free to move vertically; therefore the component of the force on G along the inclined plane FA will bear to the maximum and total force on this same body G along the perpendicular FC the same ratio as the weight in H to the weight in G. This ratio is, by construction, the same that the height, $|FC|$, of the inclined plane bears to the length $|FA|$:

$$w_{G,\,effective} = w_H = w_G \frac{|FC|}{|FA|} = w_G \sin\theta, \qquad (24)$$

(where θ is the angle formed by the inclined plane with the horizontal), and the body will move with acceleration proportional to $\sin\theta$. This is the lemma that I wanted to demonstrate and that, as you will see, was taken up by our author in his treatise.

---

[d] I have reproduced here the terms used by Galilei, which show some confusion between the concepts (the formalization of mechanics will start only from Newton).

Sagr. From what you have shown me so far, it seems to me that it can be inferred that the forces on the same body in moving on inclined planes differently, but with the same vertical height, like |FA| and |FI|, are mutually inversely proportional to the lengths of the planes.

Salv. I will now resume the reading of the text.

Proposition (theorem). *Neglecting the friction of the plane and the resistance of the air, if a body falls freely on inclined planes of any angle, but with the same height, the speeds with which it reaches the bottom are the same.*

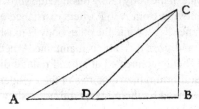

From the expressions $v = at$ *(20) and* $s = \frac{1}{2} a t^2$ *(23) we have*

$$v(s) = \sqrt{2as}. \tag{25}$$

If we call g *the acceleration in a motion along the vertical, the acceleration along a generic inclined plane of an angle* θ *with respect to the horizontal is, as we have seen,*[e]

$$a = g \sin \theta. \tag{26}$$

Thus

$$v_f = \sqrt{2as} = \sqrt{2g \sin \theta \frac{h}{\sin \theta}} = \sqrt{2gh},$$

where h *is the height AC. Quod erat demonstrandum.*

Proposition (theorem). *If the same body, starting from rest, falls along inclined planes each having the same height, or along the vertical, the times of descent will be to each other as the lengths of the inclined planes and the vertical.*

From $v_B = v_A$ *(* v_B *and* v_A *are respectively the velocities at the points B and A), and* $v = at$ *(where* $a = g \sin \theta$ *), we have, calling* $t_B et_A$ *respectively the arrival times in B and A,*

$$\frac{t_B}{t_A} = \frac{|CB|}{|CA|}:$$

Sagr. It seems to me that the above could have been proved clearly and briefly based on a proposition already demonstrated, namely, that the distance traversed in the case of accelerated motion along AC or AB is the same as that covered by a

---

[e] The vertical acceleration for a material point is g. For a sphere rolling the acceleration is $g' < g$, since part of the energy goes into rotational energy. The concept is, however, the same.

uniform speed whose value is half the maximum speed; the two segments AC and AB having been traversed at the same uniform speed it is evident that the times of descent will be to each other as the distances.

Corollary. *The times of descent along planes having different inclinations, but the same vertical height, stand to one another in the same ratio as the lengths of the planes.*

Proposition (theorem). *In general, for planes of length s, slope θ and height h, the time of descent is*

$$t = \sqrt{\frac{2s}{a}} = \sqrt{\frac{2h/\sin\theta}{g\sin\theta}} = \sqrt{\frac{2h}{g\sin^2\theta}}. \tag{27}$$

Proposition (theorem). *If from the highest or lowest point of a vertical circle one draws inclined chords that meet the circumference, the times of descent along each of them are equal.*

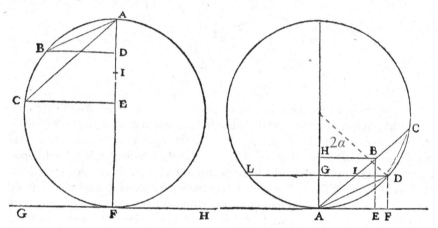

Calling *R* the radius of the circle, let *s* be the length of a generic chord. The time of descent along this chord (Eq. 23) will be such that

$$t^2 = \frac{2s}{a}.$$

If 2α is the angle at the center of a generic chord AD, the elevation of the inclined plane is α, and the length of the chord is s = 2R sin α. We have

$$t^2 = \frac{2s}{a} = \frac{4R\sin\alpha}{g\sin\alpha} = \frac{4R}{g},$$

and therefore the descent times do not depend on the chosen chord, as we wanted to demonstrate.

Corollary 1.*The descent times of a body along all the chords conducted from the extremities of the circumference to the lowest point are equal.*[f]

Corollary 2. *If from any point a vertical line and an inclined line are drawn along which the descent times are the same, the inclined line is a chord of a half-circle of which the vertical line is the diameter.*

224-
241
Sagr. Please allow me to interrupt the lecture for a moment so that I may clear up an idea that just occurs to me; one that, if it involves no fallacy, suggests at least a freakish and interesting circumstance, such as often occurs in nature and in the realm of necessary consequences.

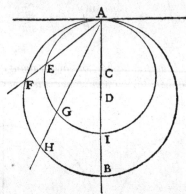

If, from any point fixed in a horizontal plane, straight lines are drawn extending indefinitely in all directions, and if we imagine a point to move along each of these lines with constant speed, all starting from the fixed point at the same instant and moving with equal speeds, then it is clear that all of these moving points will lie upon the circumference of a circle which grows larger and larger, always having the fixed point as its center. This circle spreads out in the same manner as the little waves do in the case of a pebble drop into quiet water, where the impact of the stone starts the motion in all directions, while the point of impact remains the center of these expanding circular waves.

But imagine a vertical plane from the highest point of which are drawn lines inclined at every angle and extending indefinitely; imagine also that heavy particles descend along these lines each with a naturally accelerated motion and each with a speed appropriate to the inclination of its line. If these moving particles are always visible, what will be the locus of their positions at any instant? Now the answer to this

---

[f] This is the closest demonstration to the isochronism of the pendulum oscillations, already affirmed by Galilei on the first day. Note however that the pendulum moves along the arc and not along the chord, and therefore equality is true only to the extent that we can approximate the chord with the arc, i.e. for small angles: isochronism is only an approximate property. Galilei does not make this observation, and his affirmation of the first day is incorrect for orders higher than the first in a series development of powers of the maximum angle with respect to the vertical. Equality holds for order 0 and order 1 in the angle. For a typical home pendulum clock, with a starting angle of 3 degrees (0.05 radians), the difference between the true period and the small angle approximation amounts to about 15 s per day.

question surprises me, for I am led by the preceding theorems to believe that these particles will always lie upon the circumference of a single circle, ever increasing in size as the particles recede farther and farther from the point at which their movement began. To be more definite, let A be the fixed point from which are drawn the lines AF and AH inclined at any angle. On the perpendicular AB take any two points C and D about which, as centers, circles are described passing through point A, and cutting the inclined lines at the points F, H, B, E, G, I. From the preceding theorems it is clear that, if particles start, at the same instant, from A and descend along these lines, when one is at E another will be at G and another at I; at a later instant they will be found simultaneously at F, H and B; these, and indeed an infinite number of other particles traveling along an infinite number of different slopes, will at successive instants always lie upon a single ever expanding circle. The two kinds of motion occurring in nature give rise therefore to two infinite series of circles, at once resembling and differing from each other; the one takes its rise in the center of an infinite number of concentric circles; the other has its origin in the contact, at their highest points, of an infinite number of eccentric circles; the former are produced by motions that are equal and uniform; the latter by motions that are neither uniform nor equal among themselves, but that vary from one to another according to the slope.

Further, if from the two points chosen as origins of motion, we draw lines not only along horizontal and vertical planes but in all directions, then just as in the former cases, beginning at a single point ever expanding circles are produced, so in the latter case an infinite number of spheres are produced about a single point, or rather a single sphere that expands in size without limit; and this in two ways, one with the origin at the center, the other on the surface of the spheres.

Salv. The idea is really beautiful and worthy of the clever mind of Sagredo.

Simp. As for me, I understand in a general way how the two kinds of natural motions give rise to circles and spheres; and yet as to the production of circles by accelerated motion and its proof, I did not fully understand; but the fact that one can take the origin of motion either at the inmost center or at the very top of the sphere leads one to think that there may be some great mystery hidden in these true and marvelous results, a mystery related to the creation of the Universe (that is said to be spherical in shape), and related also to the seat of its first cause.

Salv. I have no hesitation in agreeing with you. But profound considerations of this kind belong to a higher science than ours. We must be satisfied to belong to that class of less worthy workmen who procure from the quarry the marble out of which, later, the gifted sculptor produces those masterpieces that lay hidden in this rough and shapeless exterior. Now, please, let's proceed.

Corollary 4. *The descent times along all the inclined planes that intersect the same vertical circle, at its highest or lowest point, are equal to the fall time along the vertical diameter. For those planes that intersect the circumference before this diameter the times are shorter; for planes that cut this diameter, the times are longer.*

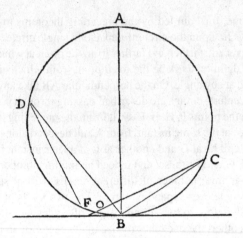

The first statement has already been proven.

The proof of the second is trivial: the DF plane is more inclined and shorter than DB, and therefore the travel time is shorter. On the contrary, CO is longer and less inclined than CB, and therefore the travel time is greater.

Proposition (theorem). *If the heights of two inclined planes stand together in the same proportion as the squares of their lengths, the bodies that start from stationary will travel these planes in equal times.*

$$t^2 = \frac{2s}{a} = \frac{2s}{gh/s} = \frac{2s^2}{gh},$$

quod erat demonstrandum.

Proposition (theorem). *The descent times along inclined planes of the same height h, but of different slopes, are between them like the lengths s of these planes; and this is true both if the movement starts from rest or if it is preceded by a fall from a height h'.*

If the body starts from rest, if the height h is constant,

$$t^2 = \frac{2s}{a} = \frac{2s}{gh/s} = \frac{2s^2}{gh} \implies t \propto s.$$

By falling from a height h' the body acquires a speed at the beginning of the motion $v_0 = \sqrt{2gh'}$, and thus

$$s = v_0 t + \frac{1}{2}g\frac{h}{s}t^2.$$

Hence

$$t = \frac{\sqrt{v_0^2 + 2as} - v_0}{a}.$$

But $2as = 2gh$ *is only a function of* h, *and is thus constant; hence*

$$t \propto \frac{1}{a} = \frac{1}{gh/s} \propto s \,.$$

Proposition (theorem). *If, after descending along an inclined plane, the movement continues along a horizontal line, the distance traveled by a body, during a time equal to the time of the fall, will be exactly double with respect to the distance in which the fall occurred.* 242-243

The average speed during the acceleration phase is, by the definition of uniformly accelerated motion,

$$< v >= \frac{v_f}{2} \,,$$

and therefore the space traveled in the fall time will be half of the space traveled after the fall, during the horizontal motion.

Comment. *The same result may be obtained by another approach. Let us consider the triangle ABC, that, by lines drawn parallel to its base, represents for us a velocity increasing in proportion to time; if these lines are infinite in number, just as the points in the line AC are infinite or as the number of instants in any interval of time is infinite, they will form the area of the triangle. Let us now suppose that the maximum velocity attained—represented by the line BC—is continued, without acceleration and at constant value through another interval of time equal to the first. From these velocities will be built up, in a similar manner, the area of the parallelogram ADBC, that is twice that of the triangle ABC; accordingly, the distance traversed with these velocities during any given interval of time will be twice that traversed with the velocities represented by the triangle during an equal interval of time. But along a horizontal plane the motion is uniform since here it experiences neither acceleration nor retardation; therefore we conclude that the distance CD traversed during a time interval equal to AC is twice the distance AC; for the latter is covered by a motion, starting from rest and increasing in speed in proportion to the parallel lines in the triangle, while the former is traversed by a motion represented by the parallel lines of the parallelogram that, being also infinite in number, yield an area twice that of the triangle.*

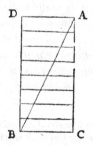

Furthermore, we may remark[g] that any velocity once imparted to a moving body will be rigidly maintained as long as the external causes of acceleration or retardation are removed, a condition that is found only on horizontal planes (in the case of planes inclined downwards there is already a cause of acceleration, while on planes sloping upward there is retardation). From this it follows that motion along a horizontal plane is perpetual: if the velocity is uniform, it cannot be diminished or slackened, much less zeroed.

244

Further, although any velocity that a body may have acquired through natural fall is permanently maintained so far as its own nature is concerned, yet it must be remembered that if, after descending along a plane inclined downwards, the body is deflected to a plane inclined upward, there is already existing in this latter plane a cause of retardation; in any such plane this same body is subject to a natural acceleration downwards. Accordingly, we have here the superposition of two different states: the velocity acquired during the preceding fall which if acting alone would carry the body at a uniform rate to infinity, and the velocity resulting from the natural acceleration downwards common to all bodies. It seems altogether reasonable, therefore, if we wish to trace the future history of a body that has descended along some inclined plane and has been deflected along some plane inclined upwards, to assume that the maximum speed acquired during descent is maintained during the ascent. In the ascent, however, the natural inclination downwards supervenes. If perhaps this discussion is a little obscure, the following figure will help to make it clearer.

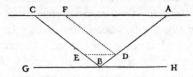

Let us suppose that the descent took place along the downward slope AB, from which the body is deflected so as to continue its motion along the upward sloping plane BC; and first let these planes be of equal length and placed so as to make equal angles with the horizontal line GH. Now it is well known that a body, starting from rest at A, and descending along AB, acquires a speed that is proportional to time, that is a maximum at B, and that is maintained by the body so long as all causes of acceleration or retardation are removed; the acceleration to which I refer is that to which the body would be subject if its motion were continued along the plane AB extended, while the retardation is that which the body would encounter if its motion were deflected along the plane BC inclined upwards; but, upon the horizontal plane GH, the body would maintain a uniform velocity equal to that which it had acquired at B after fall from A; moreover, this velocity is such that, during an interval of time equal to the time of descent through AB, the body will traverse a horizontal distance

---

[g] Here what is now called the principle of inertia is explicitly presented for the first time in this text. It is a paradigm shift for physics: before Galilei, it was believed that once the cause of motion was lacking, a body would slow down until it stopped. A mere detailed discussion will be done in the additional day.

equal to twice AB. Now let us imagine this same body to move with the same uniform speed along the plane BC so that here also during a time interval equal to that of descent along AB, it will traverse the extension of BC by a distance twice AB. But let us now suppose that when the body begins its ascent it is subjected, by its very nature, to the same influences that surrounded it during its descent from A along AB, namely, it descends from rest under the same acceleration as that which was effective in AB, and it traverses, during an equal interval of time, the same distance along this second plane as it did along AB. It is clear that, by superposing upon the body a uniform motion of ascent and an accelerated motion of descent, it will be carried along the plane BC as far as the point C where these two velocities become equal.

If now we assume any two points D and E, equally distant from the vertex B, we may then infer that the descent along BD takes place in the same time as the ascent along BE. Draw DF parallel to BC; we know that, after descent along AD, the body will ascend along DF; or, if, on reaching D, the body is carried along the horizontal DE, it will reach E with the same impetus with which it left D; hence from E the body will ascend as far as C, proving that the velocity at E is the same as that at D.

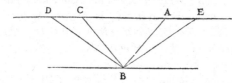

245-
262

From this we may logically infer that a body that descends along any inclined plane and continues its motion along a plane inclined upwards will, on account of the impetus acquired, ascend to an equal height above the horizontal; so that if the descent is along AB the body will be carried up the plane BC as far as the horizontal line ACD: and this is true whether the inclinations of the planes are the same or different, as in the case of the planes AB and BD. But we previously saw that the speeds acquired by fall along variously inclined planes having the same vertical height are the same. If therefore the planes EB and BD have the same slope, the descent along EB will drive the body along BD as far as D; and since this propulsion comes from the speed acquired reaching point B, it follows that this speed at B is the same whether the body has made its descent along AB or EB. Then the body will be carried up BD whether the descent has been made along AB or along EB. The time of ascent along BD is however greater than that along BC, just as the descent along EB occupies more time than that along AB; moreover it has been demonstrated that the ratio between these times is the same as that between the lengths of the planes.

Proposition (theorem). If from the lowest point of a vertical circle an inclined plane DC is traced which subtends an arc smaller than a quadrant, and if from the end of this plane we conduct two other planes DB and BC which touch the arc at any point B included between D and C, the descent time along the sequence of these two planes will be shorter than the motion along the DC plane, and also of the motion along the lower plane only.

263-
266

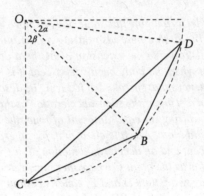

We want to prove that

$$t_{DC} - t_{DBC} = (t_{DC} - t_{DB}) - t_{BC}(v_B) > 0 \tag{28}$$

where t_{DC} and t_{DB} are respectively the times necessary for a particle to travel the DC and DB sections starting from rest, t_{DBC} is the time required to travel first the section DB and then the section DC starting from rest in D, and $t_{BC}(v_B)$ is the time needed to travel the section BC by starting the descent with a speed v_B.

The elevation angle of the plane DC with respect to the horizontal is given by $\alpha + \beta \equiv \varphi$, while the elevation angle of DB is $\varphi + \beta$; $D\widehat{C}B = \alpha$ and $B\widehat{D}C = \beta$.

Since the body starts from rest in D

$$v_C = \sqrt{2g\,|DC|\sin\varphi}\ ;\ \ v_B = \sqrt{2g\,|DB|\sin(\varphi+\beta)}$$

(note that, as previously demonstrated, v_C does not depend on whether the body has descended from D to C along a segment or along a polygonal), and

$$t_{DC} = \sqrt{\frac{2|DC|}{g\sin\varphi}}\ ;\ \ t_{DB} = \sqrt{\frac{2|DB|}{g\sin(\varphi+\beta)}}\ .$$

Thus

$$t_{DC} - t_{DB} = \frac{\tilde{v}_C - \tilde{v}_B}{g\sqrt{\sin\varphi\sin(\varphi+\beta)}} \tag{29}$$

with

$$\tilde{v}_C = \sqrt{g\,|DC|\sin(\varphi+\beta)}\ ;\ \ \tilde{v}_B = \sqrt{g\,|DB|\sin\varphi}\ .$$

From (29) we obtain, with an algebraic manipulation,

$$t_{DC} - t_{DB} = \frac{2|BC|\sin 2\varphi}{v_C\sin(\varphi+\beta) + v_B\sin\varphi}\ . \tag{30}$$

Since

$$t_{BC}(v_B) = \frac{|BC|}{(v_C + v_B)/2},$$

we obtain

$$t_{DC} - t_{DBC} = \frac{2|BC|}{(v_C + v_B)(v_C \sin(\phi + \beta) + v_B \sin\phi)} \cdot \Delta, \qquad (31)$$

with

$$\Delta = v_C(\sin 2\phi - \sin(\phi + \beta)) + v_B(\sin 2\phi - \sin\phi)$$
$$= 2v_C \sin\frac{\alpha}{2} \cos\frac{3\phi + \beta}{2} + 2v_B \sin\frac{\phi}{2} \cos\frac{3\phi}{2}.$$

Note that $2\alpha + 2\beta = 2\phi < \pi/2$, $\alpha > 0$ $\beta > 0$, thus the expression (31) is always positive. The first statement in the theorem thesis is therefore demonstrated.

The second statement is trivial, being, as previously demonstrated, $t_{BC} = t_{DC}$ (if the body starts with zero speed in both cases).

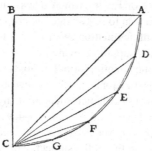

From the previous demonstration it seems possible to conclude that the fastest movement between two points does not take place along the shortest line, that is, along a segment of a straight line, but along an arc of a circle.[h] In the quadrant BAEC, having the side BC vertical, divide the arc AC into any number of equal parts, AD, DE, EF, FG, GC, and from C draw straight lines to the points A, D, E, F, G; draw also the straight lines AD, DE, EF, FG, GC. Evidently, the descent along the path ADC is quicker than along AC alone or along DC from rest at D. But a body, starting from rest at A, will traverse DC more quickly than the path ADC; while, if it

---

[h] The demonstration that follows is not correct, and in fact the conclusion is false: the line along which the motion is fastest, the so-called *brachistochrone*, is not an arc of a circumference. The problem of the brachistochrone was solved only in 1697 by Johan Bernoulli, using techniques of differential calculus that were not known in Galilei's time: the brachistochrone curve is the cycloid. In addition, there is also a logical fallacy in Galilei's reasoning which leads to the demonstration of the (true) fact that motion along the circumference is faster than motion along any inscribed polygon. This is a "venial" fallacy: from the elements presented the conclusion could be correctly deduced, but Galilei forgets to demonstrate a necessary proposition, that is, that the descent time along two planes is shorter than the motion along a single plane even when a body starts from the top with the speed that would come from its previous fall.

starts from rest at A, it will traverse the path DEC in a shorter time than DC alone. Hence the descent along the chords, ADEC, will take less time than along the two chords ADC. Similarly, the time required to traverse EFC is less than that needed for EC alone. Therefore the descent is more rapid along the four chords ADEFC than along the three ADEC. And finally a body, after descending along ADEF, will traverse the two chords, FGC, more quickly than FC alone. Therefore, along the five chords, ADEFGC, the descent will be more rapid than along the four, ADEFC. Consequently the nearer the inscribed polygon approaches a circle the shorter is the time required for descent from A to C. What has been proven for the quadrant holds true also for smaller arcs; the reasoning is the same.

Sagr. Indeed, I think we may concede to our Academician, without flattery, his claim that in the principle on accelerated motion laid down in this treatise he has established a new science dealing with a very old subject. Observing with what ease and clearness he deduces from a single principle the proofs of so many theorems, I wonder not a little how such a question escaped the attention of Archimedes, Apollonius, Euclid and so many other mathematicians and illustrious philosophers, especially since so many ponderous tomes have been devoted to the subject of motion.

267   Salv. There is a fragment of Euclid which deals with motion, but there is no indication in it that he ever began to investigate the property of acceleration and the manner in which it varies with inclination. We may say the door is now opened, for the first time, to a new method supported by numerous and wonderful results that in future years will command the attention of other minds.

Sagr. I really believe that just as, for instance, the few properties of the circle proven by Euclid in the Third Book of his Elements lead to many others more recondite, so the principles that are outlined in this little treatise will, when taken up by speculative minds, lead to many another more remarkable result; and it is to be believed that it will be so on account of the nobility of the subject, which is superior to any other in nature.

During this long and laborious day, I have enjoyed these simple theorems more than their proofs, many of which, for their complete comprehension, would require more than an hour each. This study, if you will be kind and leave the book in my hands, is one which I mean to discuss after we have read the remaining portion that deals with the motion of projectiles; and this, if it is acceptable to you, we shall take up tomorrow.

Salv. I shall not fail to be with you!

The third day ends.

Day Four

(The Motion of Projectiles)

Salviati. Simplicio is here on time; so let us without delay take up the question of motion. The text of our Author follows:

ON THE MOTION OF PROJECTILES

We have discussed the properties of uniform motion and of naturally accelerated motion along planes of different inclinations. I now propose to set forth those properties that belong to a body whose motion is composed of two motions, namely, one uniform and one naturally accelerated; I propose to demonstrate formally these properties, well worth knowing. This is the kind of motion seen in a moving projectile.

Imagine a particle projected along a horizontal plane without friction; then we know, from what has been explained in the preceding pages, that this particle will move along this same plane with a motion that is uniform and perpetual, provided the plane has no limits. But if the plane is limited and elevated, then the moving particle, that we imagine being a heavy one, passing over the edge of the plane will acquire, in addition to its previous uniform and perpetual motion, a downward propensity due to its own weight; so that the resulting motion, that I call projection, is composed of one that is uniform and horizontal and of another that is vertical and naturally accelerated. We now proceed to demonstrate some of its properties, the first of which is as follows:

Proposition (theorem). A projectile that moves by a uniform horizontal motion *composed with a naturally accelerated vertical motion describes a semi-parabolic path.*

Sagredo. Here, Salviati, it will be necessary to stop a little while for my sake and, I believe, also for the benefit of Simplicio. I have not gone very far in my study of Apollonius and am merely aware of the fact that he discusses the parabola and other conic sections,[a] I hardly think one will be able to follow the proof of other propositions depending upon the original ones. Since even in this first beautiful theorem the author finds it necessary to prove that the path of a projectile is a parabola, and since, as I imagine, we shall deal with only this kind of curves, it will be absolutely necessary to have a thorough acquaintance, if not with all the properties that Apollonius has demonstrated[38] for these figures, at least with those which are needed for the present discussion.

Salv. Don't be so shy. At the time when we were discussing the strength of materials, we needed to use a certain theorem due to Apollonius, and that gave you no trouble.

Sagr. I may possibly have assumed it, as long as needed, for that discussion; but now when we have to follow all these demonstrations about such curves I ought not to swallow it without the possibility to digest it, thus wasting time and energy.

Simplicio. Sagredo is, as I believe, well equipped for all needs while I do not understand even the elementary terms. Although our philosophers have treated the motion of projectiles, I do not recall their having described the path of a projectile except to state in a general way that it is always a curved line, unless the projection be vertically upwards. But if the little Euclid that I have learned since our previous discussion does not enable me to understand the demonstrations that are to follow, then I shall be obliged to accept the theorems on faith without fully understanding them.

Salv. On the contrary, I desire that you should understand them from the Author himself, who, when he allowed me to see this work of his, was good enough to prove for me two of the principal properties of the parabola[39] because I did not happen to have at hand the books by Apollonius.

Referring to the figure, suppose that a plane cuts a straight cone parallel to the side to which the *lκ* segment belongs; a parabola is defined as the section traced on the surface of the cone by the plane itself (to which the *bac* section belongs).

270-
271

---

[a] The term "conic section" indicates a curve represented in a Cartesian plane by a second-degree equation (parabola, hyperbola, ellipse with the particular case of the circumference). Before the advent of analytical geometry, born with Descartes and therefore contemporary to Galilei, these curves were built on the basis of geometric properties, and in particular by sectioning a straight circular cone with a plane. Intersecting the cone with a plane so as to form a closed curve this curve is an ellipse (in particular a circumference if the plane is perpendicular to the axis of the cone). By sectioning it with a plane parallel to one of the generating lines, a parabola is obtained. The open curve obtained by intersecting it with a more angled plane than that the one that cuts a parabola is a branch of hyperbola.

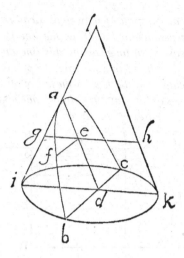

The property of the parabola that we want to demonstrate is the following:

$$\frac{|da|}{|ea|} = \frac{|bd|^2}{|fe|^2}.$$ (33)

Since bd is perpendicular to the diameter $i\kappa$ of the circle $ib\kappa$,

$$|bd|^2 = |id| \cdot |d\kappa|.$$

Similarly, for the upper circle,

$$|fe|^2 = |ge| \cdot |eh|.$$

But since $|d\kappa| = |eh|$ (the secant plane is parallel to the $l\kappa$ axis), dividing the first of the two equations by the second one has

$$\frac{|da|}{|es|} = \frac{|bd|^2}{|fe|^2}.$$

The condition (33) can be written, referring to a parabola with the vertex at point $(0,0)$ in a plane xy:

$$y = kx^2.$$ (34)

The above equation can be taken as a definition of parabola.

Now we can go back to the text and see how the Author demonstrates that a body falling with a movement composed of a uniform horizontal rectilinear motion and a vertical fall describes a semi-parabola.

272

Let us imagine an elevated horizontal line or plane ab along which a body moves with uniform speed from a to b. Suppose this plane ends abruptly at b; then at this point the body will, on account of its weight, acquire also a natural motion downwards along the perpendicular bn. Draw the line be along the plane ba to represent the flow, or measure, of time; divide this line into several segments be, cd, de, representing equal intervals of time; from the points b, c, d, e, let fall lines parallel to the perpendicular. On the first of these lay off any distance ci, on the second a distance four times as long, df; on the third, one nine times as long, eh; and so on, in proportion to the squares of cb, db, eb.

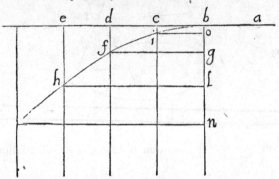

By calling x the horizontal coordinate oriented towards the left, y the height relative to the plane, and v_0 the horizontal speed, we have

$$x = v_0 t \implies t = \frac{x}{v_0}$$

$$y = -\frac{g}{2}t^2 ,$$

and therefore, by replacing in the second equation the expression of time obtained from the first,

$$y = -\left(\frac{g}{2v_0^2}\right) x^2 \tag{35}$$

that, as we wanted to demonstrate, represents a half-parabola.

273-
274
    Sagr. One cannot deny that the argument is new, subtle and conclusive, resting as it does upon the hypothesis that the horizontal motion remains uniform, that the vertical motion continues to be accelerated downwards in proportion to the square of the time, and that such motions and velocities combine without altering, disturbing, or hindering each other,[b] so that as the motion proceeds the path of the projectile

---

[b] For the first time it is made clear that one can combine a horizontal and a vertical motion treating them independently and then superimposing them, and that this property is not evident, but it tells us something about the characteristics of the phenomenon described.
    The possibility to separate independent motions is an example of the so-called *superposition principle,* granting the possibility of breaking down a problem in smaller components and then combining the solutions. Galilei first understands that this property is not obvious and must be investigated experimentally; in a similar kinematic analysis of plane motion Nicolas Oresme, in his work of 1377 *Le livre du ciel et du monde* (probably the first treatise to deal formally with the

does not change into a different curve. But this, in my opinion, is impossible. The axis of the parabola along which we imagine the natural motion of a falling body to take place stands perpendicular to a horizontal surface and ends at the center of the Earth; and since the parabola deviates more and more from its axis no projectile can ever reach the center of the Earth or, if it does, as seems necessary, then the path of the projectile must transform itself into some other curve very different from the parabola.

Simp. To these difficulties, I may add others. One of these is that we suppose the horizontal plane, which slopes neither up nor down, to be represented by a straight line as if each point on this line were equally distant from the center, which is not the case; for as one starts from the middle of the line and goes toward either end, he departs farther and farther from the center of the Earth and is therefore constantly going uphill. Whence it follows that the motion cannot remain uniform through any distance whatever, but must continually diminish. Besides, I do not see how it is possible to avoid the resistance of the medium that must destroy the uniformity of the horizontal motion and change the law of acceleration of falling bodies. These difficulties make it highly improbable that a result derived from such unreliable hypotheses should hold true in practice.

Salv. All these difficulties and objections are so well founded that it is impossible to remove them; and, as for me, I am ready to admit them all, which indeed I think our Author would also do. But, on the other hand, I ask you to consider that other eminent scientists have accepted assumptions even if not strictly true. The authority of Archimedes alone will satisfy everybody. In his treaties about mechanics and in his first quadrature of the parabola he takes for granted that the beam of a balance or steelyard is a straight line, every point of which is equidistant from the common center of all heavy bodies, and that the cords by which heavy bodies are suspended are parallel to each other.

Some consider this assumption permissible because, in practice, our instruments *275* and the distances involved are so small in comparison with the enormous distance from the center of the Earth that we may consider a minute of arc on a great circle as a straight line, and may regard the perpendiculars from its two extremities as parallel. Actually one had to consider such small quantities, it would be necessary first of all to criticize the architects who presume, by use of a plumbline, to erect high towers with parallel sides. I may add that, in all their discussions, Archimedes and the others considered themselves as located at an infinite distance from the center of the Earth, in which case their conclusions were absolutely correct. When we wish to apply our proven conclusions to distances that, though finite, are very large, it is necessary for us to infer, on the basis of demonstrated truth, what correction is to be made for the fact that our distance from the center of the Earth is not really infinite, but just very large in comparison with the small dimensions of our apparatus. The largest of these will be the range of our projectiles—and even here we need to consider only the artillery—which, however great, will never exceed four of those miles of which

---

combination of motions) had taken it for granted. In modern terms, we say that this property, that is at the basis of the reductionism, is characteristic of systems described by linear equations.

as many thousand separate us from the center of the Earth.[40] And since these paths terminate upon the surface of the Earth only very slight changes can take place in their parabolic figure that, it is conceded, would be greatly altered if they terminated at the center of the Earth.

As to the perturbation arising from the resistance of the medium, this is more considerable and cannot, on account of its manifold forms, be exactly described. Thus if we consider only the resistance that the air offers to the motions studied by us, we shall see that it disturbs them all in an infinite variety of ways corresponding to the infinite variety in the form, weight, and velocity of the projectiles. As to velocity, the greater this is, the greater will be the resistance offered by the air; the resistance will be greater as the moving bodies become less dense. So that although the falling body ought to be displaced in proportion to the square of the duration of its motion, no matter how heavy the body, if it falls from a very considerable height, the resistance of the air will be such as to prevent any increase in speed and will render the motion uniform. In proportion as the moving body is less dense, this uniformity will be more quickly attained and after a shorter fall. Even horizontal motion that, if no impediment were offered, would be uniform and constant is altered by the resistance of the air and finally ceases; and here again the less dense the body the quicker the process. Of these properties of weight, of velocity, and also of form it is not possible to give an exact description; hence, to handle this matter scientifically, it is necessary to cut loose from these difficulties, and having discovered and demonstrated the theorems in the case of no resistance, to use them and apply them with such limitations as experience will teach. The advantage of this method will not be small; for the material and shape of the projectile may be chosen as dense and round as possible, so that it will encounter the least resistance in the medium. Nor will the spaces and velocities in general be so great but that we shall be easily able to correct them with precision.

In the case of those projectiles that we use, made of dense material and round in shape, or of lighter material and cylindrical in shape, such as arrows, thrown from a sling or crossbow, the deviation from an exact parabolic path is quite insensible. Indeed, if you will allow me a little greater liberty, I can show you, by two experiments, that the dimensions of our apparatus are so small that these external and incidental resistances, among which that of the medium is the most considerable, are scarcely observable.

I now proceed to the consideration of motions through air, since it is with these that we are now especially concerned; the resistance of air exhibits itself in two ways: first by offering greater impedance to less dense than to very dense bodies, and secondly by offering greater resistance to a body in rapid motion than to the same body in slow motion.

Regarding the first of these, consider the case of two balls having the same dimensions, but one weighing ten or twelve times as much as the other; one, say, of lead, the other of oak, both allowed to fall from an elevation of 150 or 200 cubits. The experiment shows that they will reach ground with slight difference in speed, showing us that in both cases the retardation caused by air is small; if both balls start at the same moment and at the same elevation, and if the heavier one be slightly retarded and the lighter one greatly retarded, then the former ought to reach ground a

considerable distance in advance of the latter, since it is ten times as heavy. But this   *277*
does not happen; indeed, the gain in distance of one over the other does not amount
to the hundredth part of the entire fall. And in the case of a ball of stone weighing
only a third or half as much as one of lead, the difference in their times of descent
will be scarcely noticeable. Now since the speed acquired by a leaden ball in falling
from a height of 200 cubits is so great that if the motion remained uniform the ball
would, in an interval of time equal to that of the fall, traverse 400 cubits, and since
this speed is so considerable in comparison with those which, by use of bows or other
machines except fire weapons, we are able to give to our projectiles, it follows that
we may, without sensible error, regard as absolutely true those propositions that we
are about to prove without considering the resistance of the medium.

Passing now to the second case, where we have to show that the resistance of
the air for a rapidly moving body is not very much greater than for one moving
slowly, ample proof is given by the following experiment. Attach to two threads of
equal length—say four or five yards—two equal lead balls and suspend them from
the ceiling; now pull them aside from the perpendicular, the one through 80 or more
degrees, the other through no more than four or five degrees; so that, when set free, the
one falls, passes through the perpendicular, and describes large but slowly decreasing
arcs of 160, 150, 140 degrees, etc.; the other swinging through small and also slowly
diminishing arcs of 10, 8, 6, degrees, etc. In the first place, it must be remarked that
one pendulum passes through its arcs of 180°, 160°, etc., in the same time that the
other swings through its 10°, 8°, etc., from which it follows that the speed of the
first ball is 16 and 18 times greater than that of the second. Accordingly, if air offers   *278*
more resistance to the high speed than to the low, the frequency of vibration in the
larger arcs, ought to be less than in the small arcs; but this prediction is not verified
by experiment; because if two persons start to count the vibrations, the one the large,
the other the small, they will discover that after counting tens and even hundreds
they will not differ by a single vibration, not even by a fraction of one.

Sagr. Since we cannot deny that air hinders both of these motions, both becoming
slower and finally vanishing, we have to admit that the retardation occurs in the same
proportion in each case. But how? How, indeed, could the resistance offered to the
one body be greater than that offered to the other except by the transfer of more
impetus and speed to the fast body than to the slow? And if this is so the speed with
which a body moves is at once the cause and measure of the resistance which it
meets. Therefore, all motions, fast or slow, are hindered and diminished in the same
proportion; a result, it seems to me, of no small importance.

Salv. We are able, therefore, in this second case, to say that the errors (neglecting
the accidental ones) in the results that we are about to demonstrate are small in
the case of our machines, where the velocities employed are mostly large and the
distances negligible in comparison with the semi-diameter of the Earth or one of its
great circles.

Simp. I would like to hear your reason for putting the projectiles of fire weapons,
i.e., those using powder, in a different class from the projectiles employed in bows,
slings, and crossbows, on the ground of their not being equally subject to change and
resistance from air.

Salv. I am led to this view by the excessive and, so to speak, supernatural[c] violence with which such projectiles are launched; indeed, one might say that the speed of a ball fired either from a musket or from a piece of ordnance is supernatural. If such a ball will be allowed to fall from some great elevation its speed will, owing to the resistance of air, not go on increasing indefinitely. That which happens to bodies of small density in falling through short distances—I mean the reduction of their motion to uniformity—will also happen to a ball of iron or lead after it has fallen a few thousand cubits; this terminal or final speed is the maximum that such a heavy body can naturally acquire in falling through air. This speed I estimate to be much smaller than that impressed to the ball by the burning powder.

279

An appropriate experiment will serve to demonstrate this fact. From a height of one hundred or more cubits fire a musket loaded with a lead bullet, vertically downwards upon a stone pavement; with the same gun shoot against a similar stone from a distance of one or two cubits, and observe which of the two balls is the more flattened. If the ball which has come from the greater elevation is found to be the less flattened of the two, this will show that the air has hindered and diminished the speed initially imparted to the bullet by the powder, and that air will not permit a bullet to acquire so great a speed, no matter from what height it falls; for if the speed impressed upon the ball by the fire does not exceed that acquired by it in falling freely then its downward blow ought to be greater rather than smaller.

I have not performed this experiment,[41] but I am convinced that a musket ball or cannon shot, falling from a height as great as you please, will not deliver so strong a blow as it would if fired into a wall only a few cubits distant, i.e., at such a short range that the splitting of the air will not be sufficient to rob the shot of that excess of supernatural violence given it by the powder.

But now let us proceed with the discussion in which the Author invites us to study the movement of a body when this movement is made up of two others; let's start from the case where the two are uniform, and occur in perpendicular directions.

280

Proposition (theorem). *When the motion of a body is the resultant of two uniform motions, one horizontal, the other vertical, the square of the resultant speed is equal to the sum of the squares of each of the two component speeds.*

Let us imagine any body urged by two uniform motions and let ab represent the vertical displacement, while bc represents the displacement that, in the same interval of time, takes place in a horizontal direction. If then the distances ab and be are traversed, during the same time interval, with uniform motions the corresponding velocities will be to each other as the distances |ab| and |be| are to each other; but the body which is urged by these two motions describes the diagonal ac, and its velocity is proportional to ac. The square of |ac| is equal to the sum of the squares of |ab| and |be|. Hence the square of the resultant speed is equal to the sum of the squares of the two speeds ab and be, quod erat demonstrandum.

---

[c] Here "supernatural" does not mean "miraculous", but simply impossible to achieve through a process of nature such as free fall.

Simp. At this point there is just one slight difficulty that needs to be cleared up. It seems to me that the conclusion just reached contradicts a previous proposition in which it is claimed that the speed of a body moving from *a* to *b* is equal to that of a body moving from *a* to *c*; while now you conclude that the speed at *c* is greater than that at *b*.

Salv. Both propositions, Simplicio, are true, yet there is a great difference between them. Here we are speaking of a body urged by a single motion which is the resultant of two uniform motions, while there we were speaking of two bodies each urged with naturally accelerated motions, one along the vertical *ab* the other along the inclined plane *ac*. Besides, the time intervals were there not supposed to be equal, that along the incline ac being greater than that along the vertical *ab*; but the motions of which we now speak, those along *ab*, *be*, *ac*, are uniform and simultaneous.

Simp. I am satisfied; please go on.

Salv. Our Author next explains what happens when a body is urged by a motion composed of one horizontal and uniform and of another vertical but naturally accelerated; from these two components results the path of a projectile, which is a parabola. The problem is to determine the speed of the projectile at each point. With this purpose in view our Author sets forth as follows the manner, or rather the method, of measuring such speed along the path which is taken by a heavy body starting from rest and falling with a naturally accelerated motion. *281-283*

A motion occurs along the vertical line ab, starting from rest in a, and on this line ab we choose an intermediate point c. Let |as| be the geometric mean between |ac| and |ab|:

$$|as| = \sqrt{|ab| \cdot |ac|}.$$

We will show that

$$\frac{v_C}{v_B} = \frac{|ac|}{|as|}.$$

From the equations of the uniformly accelerated motion $s(t) = \frac{1}{2}gt^2$ and $v = gt$,

$$\frac{v_C}{v_B} = \frac{t_C}{t_B} = \sqrt{\frac{|ac|}{|ab|}},$$

but

$$|as| = \sqrt{|ab| \cdot |ac|} \implies \sqrt{\frac{|ac|}{|ab|}} = \frac{|ac|}{|as|},$$

quod erat demonstrandum. Since $s(t) = \frac{1}{2}gt^2$, and $v(t) = gt$,

$$v(s) = \sqrt{2gs}\,.$$

The method for measuring the speed of a body along the direction of its fall is therefore clear: the speed increases proportionally to time, and therefore as the square root of the space traveled.

But before proceeding further, since this discussion has to do with motion consisting of a uniform horizontal motion and one accelerated vertically downwards (the path of a projectile, that is, a parabola), I remind the reader that I will call the "amplitude" of the semi-parabola ab, or even the "range", the length of the horizontal line cb; I call "height" the length of the ac axis of this parabola; I will call "sublimity" the length of the line such that a fall from that height would determine a speed equal to the speed horizontally. After presenting these definitions, I proceed with the demonstration.

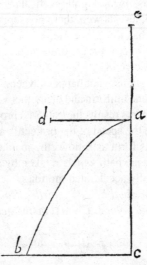

283-
285   *Sagr.* Allow me, please, to interrupt you so that I may point out the beautiful agreement between this thought of the Author and the views of Plato concerning the origin of the various uniform speeds with which the heavenly bodies revolve. The latter chanced upon the idea that a body could not pass from rest to any given speed and maintain it uniformly except by passing through all the degrees of speed intermediate between the given speed and rest. Plato thought that God, after having created the heavenly bodies, assigned them the proper and uniform speeds with which they were forever to revolve; and that He made them start from rest and move over definite distances under a natural and rectilinear acceleration such as governs the motion of terrestrial bodies. He added that once these bodies had gained their proper and permanent speed, their rectilinear motion was converted into a circular one, the only motion capable of maintaining uniformity, a motion in which the body revolves without either receding from or approaching the center. This conception is truly worthy of Plato; and it is to be all the more highly prized since its underlying principles remained hidden until discovered by our Author who removed from

them the mask and poetical dress and set forth the idea in correct perspective. Since astronomical science provides us such complete information concerning the size of the planetary orbits, the distances of these bodies from their centers of revolution, and their velocities, I cannot help thinking that our Author (to whom this idea of Plato was not unknown) had some curiosity to discover whether or not a definite "sublimity" might be assigned to each planet, such that, if it were to start from rest at this particular height and to fall with naturally accelerated motion along a straight line, and were later to change the speed thus acquired into uniform motion, the size of its orbit and its period of revolution would be those actually observed.[42]

Salv. I remember he told me that he once made the computation and found a satisfactory correspondence with observation. But he did not wish to speak of it: because of the odium that his many new discoveries had already brought upon him, this might be adding fuel to the fire. But if anyone desires such information he can obtain it for himself from the theory outlined in the present treatment. We now proceed with the matter in hand:

Proposition (problem). *Determine the speed of a bullet at each point of its parabolic path.*

The square of the speed is equal to the square of the horizontal speed, which is constant, plus the square of the vertical speed, which is proportional to the square of time, and then to the square root of the space traveled along the vertical. Using the previous relation, we have

$$v = \sqrt{v_0^2 + 2gh}\,.$$

Sagr. The way you combine these different motions to get the result affects me as so new that my mind gets confused. I am not referring to the composition of two uniform motions, even when unequal, and when one takes place along a horizontal axis, the other along a perpendicular direction, because in this case I am absolutely convinced that the result is a movement whose square is equal to the sum of the squares of the two components. Confusion arises when you combine a uniform horizontal movement with a vertical one that is naturally accelerated. I therefore trust that we can continue this discussion longer. Even in the case of the two uniform movements, one horizontal, the other perpendicular, I want to better understand how the result is obtained from the components. I hope you understand what I need.

Simp. And I feel even more the need for a clarification.

Sagr. Your request is altogether reasonable and I will see whether my long consideration of these matters will enable me to make them clear to you. But you must excuse me if in the explanation I repeat many things already said by the Author.

Concerning motions and their velocities, whether uniform or naturally accelerated, one cannot speak definitely until he has established a measure for such velocities and also for time. As for time already widely adopted hours, first minutes and second minutes. So for velocities, just as for intervals of time, there is need of a common standard which shall be understood and accepted by everyone, and which shall be the same for all. As has already been stated, the Author considers the velocity of a freely falling body adapted to this purpose, since this velocity increases according to the same law in all parts of the world; thus for instance the speed acquired by a lead ball weighting a pound starting from rest and falling vertically through the height of, say, a spear's length is the same in all places; it is therefore excellently adapted for representing the speed acquired in the case of natural fall.

It still remains for us to discover a method of measuring velocity in the case of uniform motion in such a way that all who discuss the subject will form the same conception of its size and velocity. This will prevent one person from imagining it larger, another smaller, than it really is; so that in the composition of a given uniform motion with one which is accelerated different people may not obtain different values for the resultant. In order to determine and represent such a particular speed our Author has found no better method than to use the velocity acquired by a body in naturally accelerated motion. The speed of a body which has in this manner acquired any momentum whatever will, when converted into uniform motion, retain precisely such a speed as, during a time interval equal to that of the fall, will carry the body through a distance equal to twice that of the fall. But since this matter is fundamental in our discussion, we must make it clear using a particular example.

287-
291

Let us consider the speed acquired by a body falling through the height, say, of a spear as a standard which we may use in the measurement of other speeds and momenta as occasion demands; assume for instance that the time of such a fall is four seconds.[d] Now in order to measure the speed acquired from a fall through any other height, whether greater or less, one must not conclude that these speeds bear to one another the same ratio as the heights of fall; for instance, it is not true that a fall through four times a given height confers a speed four times as great as that acquired by descent through the given height; because the speed of a naturally accelerated motion does not vary in proportion to the distance. As has been shown above, the ratio of the spaces is equal to the square of the ratio of the times.

If, then, as is often done for the sake of brevity, we take the same limited straight line as the measure of the speed, and of the time, and also of the space traversed during that time, it follows that the duration of fall and the speed acquired by the same body in passing over any other distance is proportional to the square root of the distance itself:

$$v = gt = \sqrt{2gs}\,.$$

In short, with these cautions, the height $|ab|$ can be used as a measure of the different physical quantities that enter this discussion.

Having clarified this point, let us consider the velocity in the case of two compound motions.

If both the component motions are uniform and one perpendicular to the other, we have already seen that the square of the resultant is obtained by adding the squares of the components. We therefore have the following rule to obtain the speed that derives from two uniform speeds, one vertical and the other horizontal: take the square of

---

[d] This is said without pretense of correctness: the time of fall from the height of a spear is approximately 0.9 s.

each, add the squares together and extract the square root of the sum, which will be is the resulting speed of the two. Thus, in the previous example, the body which by virtue of its vertical movement would hit the horizontal plane with a speed of 3, due to its horizontal movement, would hit ac with a speed of 4; but if the body hits with a speed that is the resultant of these two, its blow will be that of a body that moves with a speed of 5.

Let us now consider a uniform horizontal motion combined with the vertical motion of a body that falls freely starting from a standstill. It is clear that the trajectory, as has been shown, is a semi-parabola, in which the speed is always increasing because the speed of the vertical component is ever growing. Therefore, to determine the speed at a given point, it is necessary first of all to fix the speed at a uniform horizontal and then, treating the body as in free fall, find the vertical speed at the desired point; the latter can only be determined by taking into account the duration of the fall, a consideration that does not enter into the composition of two uniform motions where the speed is always the same. In this case one of the component motions has an initial value of zero and increases in proportion to time; it follows that time must determine the speed at the assigned point. Finally, it remains to obtain the velocity resulting from these two components (as in the case of uniform motions): the square of the resultant is equal to the sum of the squares of the two components. But even here it is better to give an example.

In the event of a horizontal projection from a point at altitude h, from

$$v_x = v_0 \; ; \; v_y = gt$$

and

$$h = \frac{1}{2}gt^2$$

we have

$$v = \sqrt{v_x^2 + v_y^2} = \sqrt{v_0^2 + 2gh}.$$

To what has been said concerning the momenta, blows or shocks of projectiles, we must add another very important consideration: to determine the force and energy of the shock it is not sufficient to consider only the speed of the projectiles, but we must also take into account the nature and condition of the target which, in no small degree, determines the efficiency of the blow. First of all it is well known that the target suffers violence from the speed of the projectile in proportion as it partly or entirely stops the motion; because if the blow falls upon an object which yields without resistance, such a blow will be of no effect. Likewise, when one attacks his enemy with a spear and overtakes him at an instant when he is moving with equal speed there will be no blow but merely a harmless touch. If the shock hits an object that yields only in part then the blow will not have its full effect, but the damage will be in proportion to the excess of the speed of the projectile over that of the receding body; thus, for example, if the shot reaches the target with a speed of 10 while the latter recedes with a speed of 4, the effective speed will be of 6. Finally, the blow will be maximum, in so far as the projectile is concerned, when the target does not recede at all but if possible completely resists and stops the motion of the projectile. If the target should approach the projectile the shock of collision would be greater as the sum of the two speeds is greater than that of the projectile alone.[e]

Moreover it is to be observed that the amount of yielding in the target depends not only upon the quality of the material, as regards hardness, whether it be of iron, lead, wool, etc., but also upon its position. If the position is such that the shot strikes it at right angles, the momentum imparted by the blow will be maximum; but if the motion is oblique, that is to say, slanting, the blow will be weaker; and more and more so in proportion to the obliquity. No matter how hard the material of the target is, the entire effect of the shot will not be spent and stopped: the projectile will slide by and will, to some extent, continue its motion along the surface of the opposing body.

All that has been said concerning the amount of momentum in the projectile at the extremity of the parabola must be understood to refer to a blow received on a line at right angles to this parabola or along the tangent to the parabola at the given point; even though the motion has two components, one horizontal, the other vertical, neither will the momentum along the horizontal nor that upon a plane perpendicular to the horizontal be a maximum, since each of these will be received obliquely.

Sagr. Your discussion on these blows and shocks recalls to my mind a problem in mechanics, of which no author has given a solution or said anything which diminishes my astonishment or even partly relieves my mind. My difficulty and surprise consist in not being able to see whence and upon what principle is derived the energy and immense force which makes its appearance in a percussion; for instance we see the simple blow of a hammer, weighing not more than 8 or 10 pounds, overcoming resistances which, without a blow, would not yield to the weight of a body producing impetus by pressure alone, even though that body weighed many hundreds of pounds. I would like to discover a method of measuring the force of such percussion. I can hardly think it infinite, but incline rather to the view that it has its limit and can be counterbalanced and measured by other forces, such as weights, or by levers or screws or other mechanical instruments that are used to multiply forces in a manner which I satisfactorily understand.

Salv. You are not alone in your surprise at this effect or in obscurity as to the cause of this remarkable property. I studied this matter myself for a while in vain; but my confusion merely increased until finally I met our Academician and I received from him great consolation. First, he told me that he also had for a long time been groping in the dark; but later he said that, after having spent thousands of hours in speculating and contemplating, he had arrived at some notions that are far from our earlier ideas and which are remarkable for their novelty. And since now I know that you would gladly hear what these novel ideas are I shall not wait for you to ask but promise that, as soon as our discussion of projectiles is completed, I will explain all these fantasies,

292-308

---

[e] This discussion completes that on the principle of inertia on the third day and a discussion on the principle of relativity in the *Dialogue over the two chief world systems*. The velocity addition formula described here is an example of the (today) so-called *Galilei's transformations*.

as far as I can recall them from the words of our Academician.[43] In the meantime, let's proceed with the author's propositions on the motion of projectiles.[44]

A projectile thrown at an angle ("elevation") α from the horizontal and from ground level, under the action of the combination of uniform horizontal motion and falling motion, will reach a maximum height of h, with a semi-parabola that describes the inverse of the motion described above. In h its velocity will be horizontal, and we can compute its trajectory as previously done: the body will reach the ground with another semiparabola.[45]

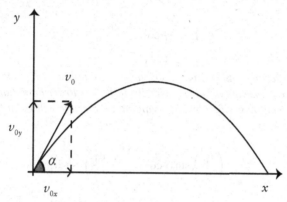

Calling x the horizontal coordinate, we will have

$$x(t) = v_{0x}t = v_0 \cos \alpha t \tag{36}$$

$$y(t) = v_{0y}t - \frac{1}{2}gt^2 = v_0 \sin \alpha t - \frac{1}{2}gt^2. \tag{37}$$

and

$$v_x(t) = v_{0x} = v_0 \cos \alpha \tag{38}$$

$$v_y(t) = v_{0y} - gt = v_0 \sin \alpha - gt. \tag{39}$$

At the maximum height $v_y = 0$, and thus[f]

$$t_{\max} = \frac{v_0 \sin \alpha}{g}$$

$$h_{\max} = y(t_{\max}) = \frac{v_0^2 \sin^2 \alpha}{2g}, \tag{40}$$

where $t_{\max}$ is the time corresponding to the maximum height. We therefore have the following relationship between launch angle, initial speed and range.

---

[f] Unlike what has been done previously, in the case of the motion of projectiles we measure angles in degrees and not in radians.

$$L = \left(\frac{v_0^2}{g}\right) 2\sin\alpha\cos\alpha = \left(\frac{v_0^2}{g}\right)\sin 2\alpha. \tag{41}$$

We have shown that the maximum range will be when the launch angle (elevation) is of 45 degrees.[46]

Let us now study the relationships between initial horizontal speed, height and range in semi-parabolic motion. From a height h, the fall time will be $t = \sqrt{2h/g}$, and therefore, calling α the angle of impact on the ground,

$$L = v_0\cos\alpha\sqrt{\frac{2h}{g}}.$$

If projectiles describe semi-parabolas of the same amplitude, the speed needed to describe the one whose amplitude is twice its altitude is lower than that required for any other. This can be verified by observing that the range will be in this case half of the range in (41):

$$L' = \left(\frac{v_0^2}{g}\right)\sin\alpha\cos\alpha = \left(\frac{v_0^2}{2g}\right)\sin 2\alpha, \tag{42}$$

where now v_0 is the speed at the impact on the ground, while the height h' will be, from (40),

$$h' = \frac{v_0^2\sin^2\alpha}{2g}. \tag{43}$$

At the angle of 45°, which minimizes v_0 for a given range,

$$L' = 2h',$$

quod erat demonstrandum.

Sagr. The force of demonstrations such as occur only in mathematics fills with wonder and delight. From accounts given by gunners, I was already aware of the fact that in the use of cannon and mortars, the maximum range, that is the one in which the shot goes farthest, is obtained when the elevation is 45° or, as they say, at the sixth point of the quadrant; but to understand why this happens far outweighs the mere information obtained by the testimony of others or even by repeated experiments.

Salv. What you say is true. The knowledge of a single fact acquired through the discovery of its causes prepares the mind to understand and ascertain other facts without the need of recourse to experiment, precisely as in the present case, where by argumentation alone the Author proves with certainty that the maximum range occurs when the elevation is 45°. He also demonstrates a fact that perhaps had never been proved before: shots that exceed a rise of 45° (mortar shots) or are lower (shots of a howitzer) by equal quantities δ have equal ranges. Indeed

$$L_{+\delta} = 2\left(\frac{v_0^2}{g}\right)\sin(90° + 2\delta) = L_{-\delta} = 2\left(\frac{v_0^2}{g}\right)\sin(90° - 2\delta).$$

Sagr. I shall be very glad to see this; from it I shall learn the difference of speed and force required to fire projectiles over the same range with extreme angles of elevation. For example, if you wanted to use a lift of 3° or 4°, or 87° or 88° and still obtain the same range that you had with an elevation of 45° (for which we have shown that the initial speed is minimal) the excess of the force required will be, I think, very large.

Salv. You are quite right; and you will see that to perform a launch at extremely high angles of elevation, you will move toward an infinite speed.

From the equation (41)

$$v_0^2 = \frac{Lg}{\sin 2\alpha}; \tag{44}$$

we first see that the statement made above is true: for different elevation angles, the greater the deviation from the optimal angle of 45 degrees and the higher the initial speed required to obtain the same range.

Sagr. I also observe that, concerning the two components of the initial speed, the higher the elevation and the greater the vertical component of the speed must be. On the other hand, when the shot reaches only a small height, the horizontal component of the initial velocity must be large.   *309*

In the case of a projectile launched with an elevation of 90°, I quite understand that all the force in the world would not be sufficient to make it deviate a single inch from the perpendicular and that it would necessarily fall back into its initial position; but in the case of zero elevation, when the shot is fired horizontally, I am not so certain that some force, less than infinite, would not carry the projectile some distance; thus not even a cannon can fire a shot in a perfectly horizontal direction, or as we say, point blank, that is, with no elevation at all. Here I admit there is some room for doubt. The fact I do not deny outright, because of another phenomenon apparently no less remarkable, but yet one for which I have conclusive evidence. This phenomenon is the impossibility of stretching a rope in such a way that it shall be at once straight and parallel to the horizon; the fact is that the cord always sags and bends and that no force is sufficient to stretch it perfectly straight.

Salv. In this case of the rope then, Sagredo, you cease to wonder at the phenomenon because you have its demonstration; but if we consider it with more care we may possibly discover some correspondence between the case of the gun and that of the string. The curvature of the path of the shot fired horizontally appears to result from two forces: one (that of the weapon) drives it horizontally and the other (its own weight) draws it vertically downward.[47] So in stretching the rope you have the force that pulls it horizontally and its own weight acting downwards. The circumstances in these two cases are, therefore, very similar. If then you attribute to the weight of the rope a power and energy sufficient to oppose and overcome any stretching force, no matter how great, why deny this power to the bullet?   *310*

Besides I must tell you something that will both surprise and please you, namely, that a rope stretched more or less tightly assumes a curve closely approximating the parabola. This similarity is clearly seen if you draw a parabolic curve on a vertical plane and then invert it so that the apex will lie at the bottom and the base remains horizontal; on hanging a chain below the base, one end attached to each extremity of the base, you will observe that, on slackening the chain more or less, it bends and resembles a parabola. The coincidence is more exact in proportion as the parabola is drawn with less curvature or, so to speak, more stretched[g]; so that using parabolas described with elevations less than 45° the chain fits its parabola almost perfectly.

Sagr. Then, with a fine chain, one could quickly draw many parabolic lines upon a plane surface.

Salv. Certainly, and with a few advantages, as I shall show you later.

Simp. But before going further, I am anxious to be convinced at least of that proposition of which you say that there is a rigid demonstration; I refer to the statement that it is impossible by any force whatever to stretch a cord so that it will lie perfectly straight and horizontal.

Sagr. I will see if I can recall the demonstration; but to understand it, Simplicio, it will be necessary for you to take for granted concerning machines what is evident not alone from experiment but also from theoretical considerations, namely, that the velocity of a moving body, even when its force is small, can overcome a very great resistance exerted by a slowly moving body, whenever the velocity of the moving body has to that of the resisting body a greater ratio than the resistance of the resisting body to the force of the moving body.

Simp. This I know very well for it has been demonstrated by Aristotle in his *Mechanical Problems.*[48] It is also clearly seen in the lever and the steelyard where a counterweight of no more than 4 pounds will lift a weight of 400 provided that the distance of the counterweight from the axis about which the steelyard rotates be more than one hundred times as large as the distance between this axis and the point of support for the large weight. This is true because the counterweight in its descent traverses a space more than one hundred times as great as that moved over by the large weight in the same time span; in other words, the small counterweight moves with a velocity that is more than one hundred times as great as that of the large weight.

Sagr. You are quite right; however small the force of the moving body, it will overcome any resistance, however great, provided it gains more in velocity than it loses in force and weight. Now let us return to the case of the rope, that I approximate as merely a line without weight.

311-312

---

[g] Again on the catenary as in the second day, but here Galilei correctly says explicitly that the catenary is just an approximation of the parabola.

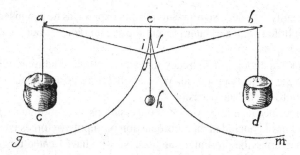

In the figure, ab represents a line passing through two fixed points, respectively a and b; at the ends of this line hang, as you can see, two large equal weights w_c and w_d, which pull it with great strength and keep it straight. If from the central point of this line, that we call e, we suspend a small weight at will, we say w_h, the line ab will drop towards the point f and, due to its lengthening, it will force the two weights in c and d to rise. This will always happen, even when w_c is much larger than w_h, and we can prove it as follows. If T is the force exerted by the wire, we will have

$$2T \sin \theta = w_h$$
$$T = w_c$$

(θ is the angle $\widehat{fae}$), from which

$$\sin \theta = \frac{w_h}{2w_c}$$

Note also that the velocity of the point h after I suspend the weight w_h fulfils the criteria that we discussed before. Indeed the movement of h is larger than the movement of c by a factor larger than the ratio between w_c and w_h:

$$\frac{|ef|}{|fi|} > \frac{w_c}{w_h}.$$

In fact

$$\frac{|ef|}{|fi|} = \frac{|ae| \tan \theta}{|ae|/\cos \theta - 1} = \frac{\sin \theta}{1 - \cos \theta},$$

and this becomes very large when the line is horizontal, as you can see from the figure.

What happens in the case of a weightless rope ab when a small weight h is attached to its central point also happens for a real rope, which, however light, still has a weight different than zero; because in this case the material of which the cable is made has the same effect as a weight suspended in its center of gravity.

Simp. I am fully satisfied. So now Salviati can explain, as he promised, the advantage of such a little chain, and later present the speculations of our Academician on the subject of impulsive forces.

Salv. I think it's enough: it's already late and the time remaining will not permit us to clear up the subjects proposed. We may therefore postpone our meeting until another and more opportune occasion.

Sagr. I agree, because after various conversations with intimate friends of our Academician I have concluded that this question of impulsive forces is very obscure, and I think that, up to the present, none of those who have treated this subject have been able to clear up its dark corners which lie almost beyond the reach of human imagination; among the various views which I have heard expressed one, strangely fantastic, remains in my memory, namely, that impulsive forces are indeterminate, if not infinite. Let us, therefore, await the convenience of Salviati.[49]

Sagr. Please, be so kind as to leave me this book until our next meeting, so that I can read and study these propositions thoroughly and in order.

Salv. I do it with great pleasure, and I hope you will enjoy reading it.

The fourth day ends.

Additional Day

(The Force of Percussion)

INTERLOCUTORS: SALVIATI, SAGREDO AND APROINO

Sagredo. Why is this new friend here and our dear Simplicio is missing?

Salviati. I imagine that some demonstrations of various problems discussed in recent days were obscure to him. This new friend you see is Paolo Aproino,[50] a noble from Treviso who was a student of our Academician during his Paduan period; and not only his pupil, but also a very close friend. With him he held long conversations together to others with whom he shared interests. Among these is the noble Daniele Antonini[51] from Udine, a man of superb intellect and value who died gloriously in defense of his country, and receiving from the great Venetian Republic honors worthy of his merit. With him, Aproino took part in a large number of experiments made at his home by our Academician, concerning a variety of problems.

About ten days ago Aproino was passing in Venice and came to see me, as he uses to do; and having heard that I had here some writings from the Author, he wanted to study them with me. Hearing about our appointment to talk about the mysterious problem of percussion, he told me he had discussed it many times with the Academician, even if in a questioning and inconclusive way, and that he was assisting at experiments related to various problems, some of which had been made about the force of percussion and its explanation. He wanted right now to mention, among others, one that he thinks to be very ingenious and subtle.

Salv. I consider it a great fortune meeting Aproino personally, since our Academician often told me about him. It will be a great pleasure for me to be able to listen to at least part of the various experiments made on different propositions, in the presence of so acute minds like that of Antonini, about whom I heard our friend speaking with praise and admiration. Now that we are here to reason specifically on the percussion, you, my dear Aproino, could tell us what was derived from the experiments on this subject. You should also promise us to speak on some other

© The Author(s), under exclusive license to Springer Nature Switzerland AG 2021
A. De Angelis, *Galileo Galilei's "Two New Sciences"*, History of Physics,
https://doi.org/10.1007/978-3-030-71952-4_5

occasions about other problems, since I know that your curiosity is inferior only to your care as an experimenter.

Aproino. Should I return your courtesy, I would have to spend so many words that very little time would remain to talk about the issue we want to discuss.

Sagr. We then leave the ceremonious compliments to the courtiers and begin immediately with a serious speech. My words are always few, but sincere.

Apr. I don't expect to say anything that Salviati doesn't already know, so the full weight of the speech should be carried by his shoulders. However, at least to begin with, I'll mention the first steps and the first experiment that our friend made to get to the heart of this admirable problem of impact. The purpose is to find and measure its great strength and, if possible, to determine at the same time its essence.

The effect of percussion seems to proceed very differently from the other mechanical effects—I say "mechanical" to exclude the immense force of gunpowder. In machines, it is evident that the speed of a weak mover can compensate for the power of a strong resistant which is moving slowly. Since in the percussion we see that the striker's movement induces a movement in the resistance, the first idea of the Academician was to try to find out which part in the impact was due to weight (for example in the case of a hammer) and which part to the speed, large or small, with which it was moved. He wanted, if possible, to measure the contribution to the percussion of each of the two components, the weight and the speed; and to achieve this, he imagined an experiment that seems to me very ingenious.

He took a very strong rod, about three cubits long, he suspended it like a balance, and put hanging at the ends of its arms two equal, very heavy weights. One of these was made of two copper buckets. The upper one, hanging from the end of the beam, was filled with water; at the handles of this bucket he hung two ropes, each about two cubits long, and to these was attached with its handles another similar bucket, but empty. At the end of the other arm, a counterweight was appended that exactly balanced the weight of the whole pair of buckets, water, and ropes at the other end. The bottom of the upper bucket had been drilled with a hole of the size of an egg or a little smaller, which could be opened and closed.[52]

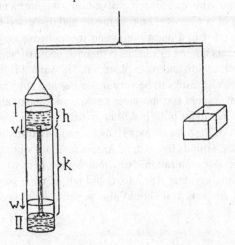

With the scale in equilibrium, the hole in the upper bucket was opened so that water could slide, and quickly descend to hit the lower bucket.

Our first conjecture was that this impact would have added some momentum on that side, so that to restore equilibrium more weight had to be added to the other arm. This addition would evidently have compensated for the strength of the impact of water, so that the force of impact of the water would have been equivalent to the weight of the ten or twelve pounds that it would have been necessary, we imagined, to add to the counterweight.

Sagr. This scheme seems really ingenious to me and I look forward to knowing the result of the experiment.

Apr. The outcome was not less surprising than we had anticipated. When the hole had been suddenly opened and the water began to fall, the scale tilted to the side with the counterweight; but when water had just begun to hit the bottom of the lower bucket the counterweight ceased to descend, and begun to rise again with a very tranquil movement, restoring balance while the water was still flowing. When equilibrium was reached, the balance stopped without moving a hairbreadth further.

Sagr. This result surprises me. It is different from what I expected, and from which I hoped to learn the entity of the force of the impact: I thought that during the fall of the water I should have increased the counterweight with an additional weight. However, it seems to me that we can derive from this outcome a lot of information.

The force of the percussion is equivalent to the weight of the quantity of water in suspension falling between the two buckets, which weighs neither on the upper bucket nor on the lower one. Not on the superior, since the parts of the water are not stuck together, and therefore cannot exert force and pull down the ones above, *325* as it would happen in some viscous liquids, like pitch. Nor does it weigh on the lower bucket, because the falling water accelerates continuously, and its upper parts do not press on the lower ones. It is as if all the water contained in the jet was not on the scale. In fact, if the falling water had a weight on the buckets, that weight together with the impact would tilt the buckets considerably downwards, raising the counterweight; and this does not happen. We have further confirmation if we imagine that all the water suddenly freezes: the jet, transformed into solid ice, would add its weight to the rest of the structure, while the cessation of movement would eliminate the force of the percussion.

Apr. We also thought like you. Furthermore, it seemed possible to conclude that the speed acquired by the fall of that amount of water from a height of two cubits, without considering the weight of this water, had exactly the same effect as the water weight, without considering the force of the impact. So if one could measure and weigh the amount of water suspended in the air between the containers, it could be seen that the impact has the same effect of a weight equal to ten or twelve pounds of falling water.

Salv. I like this smart tool, and it seems to me that without departing from that path, in which a certain ambiguity is introduced by the difficulty of measuring the quantity of falling water, we could, with a similar experiment, find a way to understand completely what we want.

Imagine, for example, one of those big weights (which I believe are called pile drivers or shins) that are used to push strong poles into the ground to build, e.g., a foundation, using the percussion caused by the fall from a certain height on such poles. Let the weight of a pile driver be, say, of 100 pounds, and let the height from which it falls be four cubits; and assume that the pole, when pushed from a single blow, will sink four inches into a hard soil. Now suppose we want to achieve the same pressure and thus the same four-inch collapse without using the impact, and we find that this can be done with a weight of 1000 pounds, which, since it only takes strength from its heaviness and without any previous movement, we can call "dead weight". I wonder if we can say that the impact of a weight of 100 pounds, combined with the speed gained in falling from a height or four cubits, is equivalent to the dead weight of 1000 pounds. That is, does the force of this speed mean as much as the pressure of 900 pounds of dead weight (which is what remains after subtracting 100 pounds of the pile driver from the 1000 pounds of the effect)?

I see that you both hesitate to answer, perhaps because I didn't explain my question correctly. To clarify things, let's suppose that the same pile driver, falling from the same height but hitting a more resistant pole, pushes it for no more than two inches. Now, can we be sure that the deadweight pressure of 1000 pounds will have this same effect? I mean, that it will drive the pole two inches?

Apr. At first thought, nobody could deny it.

Salv. And you, Sagredo, do you have any questions about it?

Sagr. Not at the moment; but having experienced a thousand times the ease with which one is deceived in reasoning, I don't feel sure of anything.

Salv. If someone like you, of whom I have known the perspicacity on many occasions, shows himself inclined to accept a wrong answer, I think it would be difficult to find even one or two people on a thousand that would not fall into such a plausible error. But you will be amazed to see how this error is hidden under a veil so thin that the slightest breeze can reveal it.

First let's let the pile driver fall on the pole as before, pushing it by four inches into the ground, and let's assume that it would take 1000 pounds to accomplish this with a dead weight. Then we bring the same pile driver to the same height, so that it falls a second time on the same pole, but now the pole sinks by only two inches, as it encountered a harder ground. Can we assume that the dead weight of 1000 pounds would cause the same effect?

Apr. So it seems to me.

Sagr. Sorry: this cannot be true. If in the first placement the dead weight of 1000 pounds drove the pole only four inches and no more, how could the pole go down another two inches simply by removing it and putting it back in the same place? Why didn't it do it before? Just by removing it and repositioning it gently can it do what it couldn't do before?

Apr. I feel ashamed: I was drowning in a glass of water.

Salv. Don't blame yourself, Aproino, because I can assure you that many others remained tied by such knots fairly easy to untie. I do not doubt that any mistake would be easy to find out if people untangled it by resolving it into its principles: at this point, something connected or close to it would clearly reveal its falsity. Our

Academician had a special genius in these cases to show with few words the absurdity and the contradiction of false statements belonging to "common sense". I have seen many conclusions related to physics that were believed to be true, and that he later proved to be false by simple reasoning.

Sagr. Indeed this is one of them, and if the others are like this, it will be good that at a certain moment you will share them with us. But, meanwhile, we continue with our question: we are looking for a way (if there is one) to assign a measure to the force of the impact. It seems to me that this cannot be obtained through the experience you proposed. Indeed, as a reasonable experiment shows us, the repeated strikes of the pile driver on the pole push it deeper and deeper, and it is clear that every next shot has an effect, which is not true for a constant dead weight. If we want dead weights to equal the effects of the third, fourth, and fifth shots, and so on, we will need bigger and bigger dead weights. Now, which of these can we take as a measure of the strength of that blow which, instead, seems always to be the same?   *328*

Salv. This is one of the main wonders that I think must have made perplexed and hesitant all speculative minds. Who, in fact, will not find it strange to hear that the measure of the force of impact should not be derived only from what is striking, but also from what receives the impact? For what concerns the mentioned experiment, it seems to me that we can deduce that the force of the impact is infinite—or rather, let us say indeterminate, or indeterminable, being now larger and now smaller, depending on whether it applies to greater or lesser resistance.

Sagr. I agree: the truth could be that the force of the impact is immense, or even infinite. In the experiment we just discussed, since the first shot plants the pole by four inches, the second by three, and as we cross ever more solid terrain, the third by two inches, and so on, repeated blows will always move the pole, but by shorter and shorter distances. But since the distance can become small at will, the penetration will continue and progress; and if we want to find a dead weight causing the same effect, each subsequent movement will require more weight than the previous one.

Salv. I can believe it.

Apr. Then there can be no resistance so great as to overcome the power of any impact, however light?

Salv. I think not, unless what is hit is completely still, that is unless its resistance is infinite.

Sagr. These statements appear to me remarkable and, so to speak, prodigious. It seems that technology can cheat physics regards to this, a sensation that at first sight we have with other mechanical tools, e.g., when a lever, a screw, or a pulley, raises huge weights with little force.   *329*

Due to the effect of percussion some hits of a hammer that weighs no more than ten or twelve pounds can flatten a copper cube that is not broken or crushed if it a marble column or even a very tall tower leans on the hammer itself. This fact seems to me to overcome any possible physical explanation. So, Salviati, please lead us out of this complicated labyrinth.

Salv. From what you say, it seems that the main problem lies in understanding how the effect of the impact, which seems infinite, can be explained by mechanisms other than those on which other machines that overcome immense resistances with very small forces are based. But I don't despair to explain this too. I will try to clarify the process; and even if it seems to me quite complicated, perhaps, thanks to your questions and objections, my observations will become thinner and sharper, and at least sufficient to widen the knot, if not to loosen it.

It is evident that in this case the roles of the motor force and of the resistance are not simple to separate. Two actions are playing a role. One is the weight, both of the mover and of the resistance; the other is the speed with which the first body moves and the other is moved. If the moved body has to move with the speed of the mover (i.e., if the spaces traversed by both in a given time are equal) it will be impossible that the weight of the mover is smaller than that of the body set in motion, but on the contrary it must be a little bigger; since in exact weight equality we would be in conditions of equilibrium, as seen in the balance of a scale with equal arms. But if with a smaller weight we want to lift a bigger one, it will be necessary to arrange the machine in such a way that the smaller moving weight crosses at the same time a greater space than the larger weight; that is to say, the former must move more quickly than the latter. And we know by experience that, for example, in the steelyard, for the counterweight to lift a weight ten or fifteen times larger, the distance between the counterweight itself and the fulcrum must be ten or fifteen times larger than the distance between the fulcrum and the suspension point of the other weight; and this means that the speed of the motor weight is ten or fifteen times greater than that of the moved weight. Since this occurs in all other machines, we can consider that at equilibrium weights and speeds are inversely proportional. We say in general, therefore, that the momentum of the weight of the lighter body balances that of the heavier one when the velocity v_m of the lighter has the same ratio with the velocity v_M of the heavier as the weight w_M has with w_m:

$$\frac{v_m}{v_M} = \frac{w_M}{w_m} \implies w_m v_m = w_M v_M .$$

If the latter is given a small advantage, the system starts moving.

Having established this, I think that not only in the percussion an action can overcome any resistance however large, as also in other mechanical devices. It is clear that a small weight of a pound will increase its effect by a factor 100 or 1000 if we place it 100 or 1000 times farther away from the fulcrum with respect to another larger weight, i.e., if the space interval through which the first descends is 100 or 1000 times larger than the space through which the other rises, so that the speed of the first is 100 or 1000 times the speed of the second.

Yet I want, by means of a more striking example, to make it clear to you that any light weight in its descent can make a very heavy mass rise high. Imagine a large weight attached to a vertical rope (a pendulum). Now imagine another thin thread hanging at the same point, and with the same length as the first; let's attach a small weight, and suppose that this small weight just touches the big one. Don't you think this new weight will push the bigger one a bit, separating its center of gravity from the vertical line where it originally lay?

Now, if this small weight can move and lift a large mass with simple contact, what will it do if it strikes it by collision? It might certainly cause such a large mass to move along the circumference.

Apr. From the results of this experiment it seems to me that the force of the impact is infinite. But this information is not enough for me to remove many dark shadows in my mind, and I still don't feel able to answer all the questions I asked myself.

Salv. Before I go any further, I want to reveal to you a possible misunderstanding. We could believe that, in the example of the pole, all the blows on the pole were equal, being made by the same pile driver always raised at the same height. But we are not allowed to draw such a conclusion.

To understand this, imagine hitting a ball that falls from above, and tell me: if, when it arrives on your hand, you let your hand descend along the same line and with the same speed as the ball, what blow would you feel? Certainly no blow at all. But if, on the arrival of the ball, you only gave in partially, causing your hand to go down at a slower speed than the ball, you would actually have an impact—not as large as that due to the entire speed of the ball, but only as the difference between its speed and the speed of your hand. So if the ball were to go down with ten degrees of speed and your hand would go down with eight, the blow would be like what you would have for a ball with two degrees of speed. If the hand went down with four degrees of speed, the blow would be like that corresponding to six degrees; and so on. You would have the entire effect of the impact only if you hold your hand steady.

Now apply this reasoning to the pile driver. The pole surrenders to the impact falling four inches on the first hit, two inches on the second, and one inch on the third. These impacts therefore have unequal effects, the first being weaker than the second and the second of the third, since the yielding of four inches takes more away from the intrinsic speed of the first percussion than what happens for the second, and the second impact is weaker than the third, which takes away twice as the speed of the second. Since the first impact is less effective than the second, and so on, it's no wonder that a lower dead weight can reproduce the first impact, and so on.

What I said explains how difficult it is to understand the strength of the impact since this acts on a resistance that varies its effect, as in the case of the pole that becomes increasingly resistant in a way that we are not able to calculate. To understand more, I think it is necessary to design an experiment in which the object that receives the impacts always opposes with the same resistance. To make an experiment of this type imagine a solid weighing, say, 1000 pounds, placed on a plane that supports it. Then, I want you to think of a rope tied to this weight that runs around a pulley at the top. It is clear that when a force is applied by pulling the end of the rope downwards, it will always meet a resistance equal to the weight to be lifted, that is, 1000 pounds of weight. If another weight equal to the first was suspended at the end of the rope, equilibrium would have been established; and if they were both lifted off the plane they would stand still.

If the first weight is supported by a plane, we can use other different weights (lower than the weight supported at rest) on the other end of the rope to test the force due to the percussion.[53] This is done by binding these weights, one at a time, to the end of the rope and then dropping them from a certain height, and observing what happens at the other end to the larger weight that feels the pull of the rope due to the other falling weight: when the rope is tightened, the tension will lift the heaviest weight with a blow that should push it upwards. It seems to me that I can predict that however small the falling weight, it should be able to overcome the resistance of the heaviest weight and lift it. This consequence seems to me definitively drawn since it is known that a smaller weight w_m prevails over another, however much greater, w_M, when

$$\frac{v_m}{v_M} > \frac{w_{\widehat{M}}}{w_m},$$

where v_m and v_M are the velocities of the minor and the major weight respectively. This always happens in the present case, since the speed of the falling weight exceeds infinitely the speed of the other, which is zero. Then we will try to find out how large the distance through which the impact received will lift the object of weight w_M, and if perhaps this distance will correspond to that of other mechanical instruments—for example in the steelyard the displacement of the greater weight is the product of the displacement of the minor weight times the ratio between the minor and the major arm. In our case we should see, assuming that the weight of the large solid initially at rest is 1000 times that of the falling weight—which falls, say, from a height of one cubit—if this raises the other weight one hundredth of a cubit; if so, it would seem to follow a rule similar to other mechanical instruments. Imagine that we are carrying out a first experiment by dropping from one height of, say, one cubit, a weight equal to the other, which we have placed on a support plane; these weights are tied to the opposite ends of the same rope. What will be the effect of the falling weight, as regards the movement and the raising of the other, who was at rest? I would be happy to hear your opinion.

Apr. Should I answer, since you look at me? It seems to me that since the two weights are equally heavy, given that the falling one has an impetus given to it by its speed, the other should be lifted well beyond the point of equilibrium, as the simple weight of the first one was enough to keep it in balance. So, in my opinion, it will go up much more than one cubit, which is the measure of the fall of the falling weight.

Salv. What do you think, Sagredo?

Sagr. At first sight the reasoning seems to me conclusive; but, as I said earlier, many experiences have taught me how easily one can be deceived and, consequently, how necessary it is to be circumspect before stating anything. So I will say, still with some doubts, that the weight of 100 pounds of the falling body will be sufficient to lift the other, which also weighs 100 pounds, up to balance. But I also think that the balance will be achieved very slowly, and therefore that when the falling body acts with great speed, it will push its partner upwards at the same speed. Now, it seems to me that more force is needed to push a heavy body upward with great speed rather than pushing it very slowly; thus it could happen that the advantage due to the speed acquired by the body in free fall along the cubit is consumed, and so to speak exhausted, in driving the other with the same speed at a similar height. I thus believe that these two movements, up and down, would end up in a state of rest immediately

after the raising weight has gone up one cubit, which would mean two cubits of   *335*
descent for the other, counting the first free fall cubit performed alone.

Salv. Since the falling weight is an aggregate of heaviness and speed, the action
of its gravity in raising the other weight is zero, being opposed by the resistance
of equal weight in the other, which clearly would not be moved without adding a
small weight. So the effect is entirely due to speed, which cannot confer anything but
speed. Not being able to confer other speed than it has acquired in the descent of one
cubit after starting from a standstill, it will push the other upwards of a similar space
and with a similar speed, consistent with what can be seen in various experiences;
that is to say, that a falling weight starting from rest has at any point sufficient speed
to bring it back to the starting height.

Sagr. This is clearly seen by examining a weight attached to a vertical cord: a
pendulum. If we move it from the vertical of an arc less than a right angle and release
it, the weight descends, and then goes back to the same angle on the other side, from
which it is evident that the ascent derives entirely from the speed acquired in the
descent, in how much the ascent cannot be due to the weight of the moving body; on
the contrary that weight, resisting the ascent, progressively removes speed.

Salv. If the example of the pendulum, of which I remember that we discussed
during the past few days, would fit well with the case we are now dealing with, your   *336*
reasoning would be very convincing. But I find significant differences between an
object that hangs from a wire, and descends from a given height along the circumference of a circle, and in the descent acquires the speed necessary to raise at an equal
height, and a weight that falls tied to the end of a rope by lifting another one, equal,
attached to the other end.

What descends along the circumference continues to gain speed up to the vertical
position thanks to its own weight, which then hinders its ascent as soon as it has
passed the vertical, the ascent being a movement contrary to heaviness. The upward
movement rewards the momentum acquired in the natural descent.

But in the other case, the falling weight pulls that other equal at rest not only
with the acquired speed but also with its heaviness; and the combination of the two
removes the resistance from the other weight to be lifted. So the previously acquired
speed does not meet opposition from the resistance of the bodies to rise, because the
rise of one body is perfectly compensated by the descent of the other.

A similar thing could happen for a heavy and round body placed on a smooth,
slightly inclined plane; this will naturally descend on its own, acquiring ever greater
speed. But if someone wanted to push it up starting from the bottom, he would need
to give it an initial speed, which with the ascent would decrease and eventually go
to zero. If the plane were not inclined, but horizontal, then this round solid will do
anything we want: if we put it at rest it will remain at rest, and if we give it a speed
in any direction, it will move in that direction, always keeping the same speed that
it will have received from our hand and without increasing or decreasing it, since
there are neither ascents nor descents on that plane. In the same way the two equal   *337*
weights, hanging at the ends of the rope, will be at rest when balanced, and if one

is given a downward speed, it will always maintain it.[a] Of course all external and accidental impediments must be removed, such as the roughness and heaviness of the rope and of the pulleys, the friction in the rotation of these around their axis, and any other resistance.

But since we are considering the speed acquired by one of these weights descending from a certain height while the other remains at rest, it will be good to determine what the speed will be with which both will be moved (one downhill and the other uphill) after the fall of one. From the demonstrations that we have already seen we know that a heavy body that falls freely starting from rest acquires an ever greater degree of speed. After the start of the traction of the other weight, this degree of speed will not increase, as its cause of increase is removed, the weight of the ascending body being the same. Then the speed will be preserved, and the movement will be converted from an accelerated motion to a uniform motion. As demonstrated and seen in the discussions of the past days, the speed will be the same as after the free fall.[b]

Sagr. Aproino reasoned better than I did. So far I am satisfied with your explanations and accept what you said to me. But probably I didn't learn enough, yet, to eliminate the wonder I feel in seeing great resistances overwhelmed by the force of the impact of a small body that strikes them even with not excessive speeds. Hearing you say that there is no finite resistance that can resist a blow without yielding increases my perplexity, as well as feeling that there is no way to assign a measure to the strength of the impact. I wish you would try to illuminate this grey area.

---

[a] The principle of inertia is affirmed here in a more general way than on the third day. Using this machine Galilei can eliminate the resultant of the forces by placing equal weights on the two sides of the pulley, and puts himself in the most general conditions—i.e., the principle of inertia is not only a property of horizontal motions, but applies to rectilinear motions in general. One is easily convinced that the opinion of some historians of science who believe, in disagreement with Newton, that Galilei is not the author of the principle of inertia, fundamental for physics, is wrong. The most famous of these historians was Alexandre Koyré [20], who thought that Galilei believed in a "circular inertia", i.e., restricted to motion around the center of the Earth. An essential element of Koyré's opinion was the denial that some of Galilei's experiments, especially those on inclined planes, were possible with the precision declared by Galilei himself. According to my opinion of experimental physicist and to those, between others, of Bellone, Drake, and Vergara Caffarelli (who had replicated these experiments with the technology presumably available to Galilei), Galilei's measures with the inclined planes are more than convincing.

It is possible however that Galilei evolved his thinking over time: in his notes of the Paduan lectures on mechanics [7] presumably dated 1598 (of which we do not possess, however, the autograph), some passages might suggest that Galilei believed in a "circular inertia". However, in the same notes Galilei explicitly writes that "all external and adventitious impediments removed, bodies can be moved in the plane of the horizon from any minimum force", a sentence that leaves little room to interpretation.

The formulation of the principle of inertia in this additional day removes all doubts, supporting the interpretation of the most authoritative of critics: Isaac Newton, who explicitly in the *Principia* attributed the principle of inertia to Galilei.

[b] Here and elsewhere in the day Galilei makes a mistake. Doubling the mass of the system in motion (the mass is a concept that Galilei did not possess: it will be introduced by Newton in the *Principia*) the speed is halved. The analysis is simple if one uses the principle of conservation of momentum, also due to Newton; in short, Galilei is often wrong by a factor of two.

Salv. No proof can be applied to a proposition unless what is assumed is certain;   *338*
and since we want to discuss the force of a striking body and the resistance of one
that receives the impact, we must choose a striker whose strength is always the same,
such as for example the same heavy body that always falls from the same height; and
similarly we establish a target of the blow that will always offer the same resistance.
To have a reproducible situation, referring to the previous example of the two heavy
bodies hanging on the ends of the same rope, I take as a striker the small weight that
is free to fall, and as the other a weight much greater than this. It is evident that the
resistance of the larger body is the same at all times and in all places, unlike what
would be the case for the resistance of a nail, or of the pole of which we spoke earlier.
For these, the resistance increases with penetration, but in an unknown way due to
the various accidental events involved, such as the hardness of the wood or the soil,
and so on, even if the nail and the pole always remain the same.

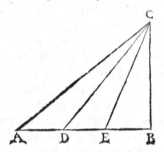

It is also necessary to recall some conclusions that we reached in the previous days
when commenting on the treatise on movement. The first is that heavy bodies, falling
from a high point to a lower horizontal plane, acquire equal degrees of speed whether
their descent is done vertically or occurs on any plane with arbitrary inclination:
what matters is only the difference between the maximum and the minimum height.
Secondly, the speed acquired by a body falling from a point C to a lower point A is the
same as that which would be needed to bring the same body back to the same height
C. From these considerations, we can understand how much force it is required to lift
that same heavy body from the horizontal axis to the C axis, whether it starts from
A, D, E or B in the figure.

Thirdly, remember that the descent times along planes between the two same   *339*
altitudes have the same ratio between them as the lengths of these planes, so that if
the plane AC, for example, is twice the length of CE and quadruple that of CB, the
descent time along CA will be double the descent time along CE and four times that
along CB.

.Finally, remember that to drag the same weight using another weight as a motor
on planes of different inclinations, the action of the weight on the steeper plane will
be more efficient and therefore it will be enough for a smaller weight, as the length
of the latter is shorter than the length of a less steep plane.

Now, assuming these truths, let's take a plane AC, say, ten times longer than the
vertical CB, and lay a solid S weighing 100 pounds on AC. It is evident that if a rope
is attached to this solid, passing over a pulley placed over point C, and a weight P of

ten pounds is fixed to the other end of this cable, with a small addition of force the weight P will go down moving the weight S along the plane AC (see the figure).

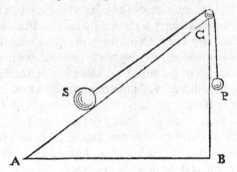

It should be noted that the distance through which the greater weight moves on the underlying plane is equal to the distance through which the small descending weight is moved; from this, someone could question the general truth that applies to all mechanical propositions, namely that a small force exceeds and moves a great resistance if the movement of the former exceeds the movement of the latter in an inverse relationship of their weights. But in the present example the descent of the small weight, which is vertical, must be compared only with the vertical ascent of the large body S, observing how much this is raised vertically from the horizon; that is, one must consider how much the distance increases along the vertical BC.

After various reflections I can affirm the following conclusion, which will then be explained and demonstrated.

340   **Proposition**. Suppose that the effect of the impact of a given weight falling from a fixed height is that of pushing a constant resistance across a certain space, and that to produce the same effect a certain amount of dead weight is needed. I say that if the original striker, acting on a greater resistance, pushes it through half the space of the first, to make this second push a double dead weight will be needed. And the same for other ratios.

In the previous example of the pole, the resistance cannot be overcome with less than a hundred pounds of dead weight. Using the force of impact, it can be overcome with a striker that weighs only ten pounds and falls from a height of, say, four cubits, and sinks the pole four inches. In the first place it is evident from what has been said above that a weight of ten pounds in vertical fall will be sufficient to lift a weight of one hundred pounds along an inclined plane with a length is ten times its height. So if the impetus acquired by the body that falls through a vertical space is applied to lift another that is equal to it in resistance, it will raise it by the same height; but the resistance of the body of ten pounds in vertical fall is equal to that of the

341   body of one hundred pounds which rises along a plane long ten times its vertical height. If the weight of ten pounds falls through any height vertically, the acquired impetus, applied to the weight of one hundred pounds, will push it for a long time corresponding to a vertical height as large as one tenth of the space on the inclined plane. And it has already been concluded that the force of the vertical weight balances a force ten times greater on this inclined plane. Thus it is evident that the vertical fall of the weight of ten pounds is sufficient to raise the weight of one hundred pounds, even vertically, but only of a space that is one tenth of that covered by the weight of ten pounds.

But that force that can lift a weight of one hundred pounds is equal to the force with which the same weight of one hundred pounds was able to push the pole into the ground when it leaned on it by pressing it. Here, then, is the explanation of how the fall of ten pounds of weight can generate a force equivalent to that of a weight of one hundred pounds, provided that the space covered by the blow is not more than a tenth of the descent of the striker. And if we now assume that the resistance of the pole has doubled or tripled, so that the pressure of two hundred or three hundred pounds of dead weight is necessary to overcome it, repeating the reasoning we will find that the impetus of the ten pounds that fall vertically is able to drive the pole the second and third time, as it did on the first time. So, by multiplying the resistance to infinity, the same blow will always be able to overcome it, but pushing the resistant body through an ever smaller space, in the inverse proportion. It seems that we might reasonably say that the impact force is infinite. But we must also consider that by changing the point of view we could consider the pressure without impact to be infinite: if this exceeds the resistance of the pole, it will continue to push it indefinitely.

Sagr. You go straight to the heart of the problem; but since it seems to me that the impact can be created in many ways and applied to a wide variety of resistances, I think it could be useful to go ahead and explain some of these practical cases, the understanding of which could open our minds to the understanding of the full subject.

Salv. I agree, and I have already thought of some examples. First, sometimes it *342* can happen that the operation of the striker becomes visible not on the struck object, but on the striker itself. If a hammer made of lead hits a fixed anvil, the effect of the blow will be seen on the hammer, which will be flattened, rather than on the anvil, which will not descend. No different from this is the effect of the mallet on the sculptor's chisel; since the mallet is made of unhardened soft iron and repeatedly hits the hardened steel chisel, it is not the chisel that is damaged, but the hammer that is dented and lacerated. Frequently we see that if you continue to hammer a nail into a very hard wood, the hammer bounces without advancing the nail. Not much different is the rebound of an inflated ball on a hard floor: the ball is deformed on impact, but soon returns to its first form, and this rebound occurs not only when what hits then recovers its own shape, but also when the same happens to what is hit: a ball bounces when it is made of very hard and inflexible material, but falls on the tightly stretched membrane of a drum.

The effect produced when a blow is added to a simple pressure, composing the two actions, is also surprising. We see it in olive presses and similar machines, when with the simple push of several men the screw is made to go down as much as possible. By retracting the bar one step, and then quickly turning it, four or six men achieve an effect that the strength of the simple push by a dozen men would not have achieved. In this case it is necessary that the bar is made of very thick and hard wood, so that it bends little or nothing; because if this were not the case, the force of the blow would be spent to twist it.[54]

343-
345

In each body to be moved by the action of a force, two different species of resistance are at work. One concerns that internal resistance which makes us say that a body weighing a thousand pounds is more difficult to lift than one weighing a hundred pounds; the other refers to the space through which the movement must be made. Two different motors correspond proportionally to these different resistances: the one that moves by pressing without striking, and the other that acts by striking. The motor that operates without impact moves a lower resistance, but it can move it for an infinite distance, always accompanying the movement with its own force.

What moves by striking moves any resistance, however large, but moves it only through a limited distance. I therefore consider these two true propositions: that a percussion can move an infinite resistance through a finite and limited interval, while a continuous force can move a finite and limited resistance through an infinite interval. These things make me doubt that Sagredo's question has an answer, just as the questions from those who try to compare things that are incommensurable—and such, I believe, are the actions of the percussion and of the pressures.

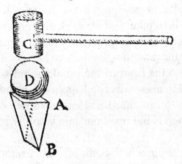

For example, in the particular case illustrated by the figure, any immense resistance in the wedge BA will be overcome by any percussion C, but only through a limited space interval, while the pressure force by a weight D will be able to overcome only a limited resistance existing in the wedge BA, and no greater than the weight D. The latter, however, will not only push through the limited interval between points B and A, but indefinitely, provided that the resistance in the body AB always remains the same, as must be assumed in the absence of other specifications.

The yielding of a material hit by a heavy body moved at any speed cannot however take place instantly, because this fact would imply an instantaneous movement through a finite space, that is clearly impossible. Therefore it takes a nonzero time time for the struck object to acquire movement from the striker.

The end.

Endnotes

1. Aristotle, *Physics*, 215a.24–216a.26.
2. Aristotle, *Physics*, 225a.25–26; *De anima*, 217a.17.
3. Aristotle, *On the Heavens*, 311b.33.
4. G. Galilei, *Discorso intorno alle cose che stanno in su l'acqua e che in quella si muovono*, Firenze 1612, IV, 106–107.
5. The authenticity of this work is doubtful; many scholars attribute it to the peripatetic school but not to Aristotle himself.
6. Giovanni de Guevara (1561–1641), bishop of Teano, had discussed this problem with Galilei and writes about it in his work *In Aristotelis mechanicas comentarii* (Rome 1627).
7. Note that Galilei approaches the concept of limit, which will be later formalized by Newton and Leibniz in particular.
8. In the text Galilei refers to the work of Luca Valerio [47]. The principle according to which if two solids have the same height and if the sections cut by planes parallel to the bases and equally distant from them are always in a given ratio, the volumes of the solids will also be in this ratio, today known as the Cavalieri principle, was demonstrated by Father Bonaventura Cavalieri (1598–1647), a pupil of Galilei, and published after the *Two New Sciences*, in 1647.
9. Added figure; adapted from Wikimedia Commons.
10. Aristotle, *Physics*, 215a.24–216a.21; *On the Heavens*, 301b.
11. Aristotle, *Physics*, 215a25.
12. G. Galilei, *Discorso intorno alle cose che stanno in su l'acqua e che in quella si muovono*, Firenze 1612, IV, 103.
13. Aristotle, *On the Heavens*, books 1 and 2.
14. Aristtele, *On the Heavens*, 311b9-10.
15. Handwritten in the original copy by Galilei.
16. Aristotle, *Mechanical Problems*, 3, 9.
17. Archimedes, *On the Equilibrium of Planes*, I, Props. 6, 7.

© The Editor(s) (if applicable) and The Author(s), under exclusive license to Springer Nature Switzerland AG 2021
A. De Angelis, *Galileo Galilei's "Two New Sciences"*,
History of Physics, https://doi.org/10.1007/978-3-030-71952-4

18. Archimedes, *On the Equilibrium of Planes*, I, Props. 6, 7.
19. Aristotle, *Mechanical Problems*, 27.
20. Ariosto, *Orlando Furioso*, XVII, 30. In the original

    Non si può compartir quanto sia lungo
    sì smisuratamente è tutto grosso.

21. Aristotle, *Mechanical Problems*, 14.
22. We omit a problem and a demonstration descending trivially from (14).
23. We slightly modified Galilei's drawing by adding a *z* axis under the prism.
24. Archimedes, *On spirals*, Prop. 10. Galileo cites the principle without repeating
    the proof; the reader·can prove it by induction, or by demonstrating the recurring
    formula
    $$1^2 + ... + n^2 = \frac{n(n + 1)(2n + 1)}{6}.$$

25. Luca Valerio, *Quadratura Parabolae*, Roma, 1606, Prop. IX.
26. This posthumous tribute to Luca Valèrio (1553-1618) seems excessive and is
    difficult to explain. Galilei had renounced the publication of his first works on the
    same topic, later included as an appendix to the first edition of the *Discourses*,
    due to Valerio's work; the two were exchanging letters in the Paduan period of
    Galilei. Valerio, however, had opposed the Copernican ideas in 1616, and the
    correspondence had ended.
27. The discussion on this problem is made explicitly by Galilei, but we have omitted
    it because it is redundant compared to the algebraic demonstration. Galilei also
    illustrates a geometric technique to solve the problem of finding the radius *R* of
    a full cylinder of equal mass with respect to a hollow tube; the algebraic solution
    is trivial:
    $$\pi R^2 = \pi (R_{ext}^2 - R_{int}^2).$$

28. This definition and the consequent Equation (20) replace axioms I-IV and the-
    orems (propositions) I-VI in the original work by Galilei.
29. The following figure, Salviati's intervention that includes it and Sagredo's inter-
    locutory answer do not appear in the original 1638 edition, but only in the second
    edition published by Dozza in Bologna in 1655 (edited by Vincenzo Viviani,
    equal to the first of 1638 apart from this detail) and in the following ones, in
    particular in that of 1718. Viviani says that the addition had been dictated to him
    by Galilei, then blind, at the end of 1638, during a re-reading of the first edition,
    and that they had revised it in November 1639. The figure we have reported
    in the text is, for homogeneity, that of the 1718 edition; for completeness we
    reproduce in this note its first appearance in 1655.

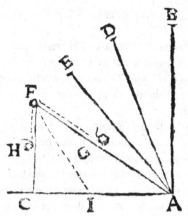

30. Galileo's lecture notes on mechanics taken from his lectures at the University of Padua were handwritten; Father Marin Mersenne published a printed French translation in 1634 [9]. The Italian publication is posthumous, and was edited by Galilei's students in 1649 [8]. However, the main theorems were included in these *Two New Sciences*.

31. Figure modified from the original.

32. This is a quotation from a sonet by Michelangelo:

   "Non ha l'ottimo artista alcun concetto
   che un marmo solo in sé non circonscriva
   col suo superchio, e solo a quello arriva
   la mano che ubbidisce all'intelletto."

   (son. 151, 1-4). In practice, any potential statue is already in the marble, and the artist with his hand guided by the intellect can remove the superfluous and discover it.

33. We have omitted before these some demonstrations and problems (propositions from 9 to 23 inclusive) that contain information included in the equations

$$s = \frac{1}{2}at^2 \; ; \; a = \frac{gh}{s}$$

   that we already demonstrated.

34. For the reason illustrated in note 40 we omitted propositions from 26 to 35 included.

35. In this demonstration we partly followed the article by R. Mandl, T. Pühringer and M. Thaler, The American Mathematical Monthly, 119, No. 6 (2012), pp. 468-476. The authors present a more complete proof, that is valid even if the arrival point is not the lowest of the circumference, and also a more elegant proof.

36. For the reason illustrated in note 40 we omitted the propositions 37 and 38.

37. The fragment is of doubtful authenticity, as it appears only in editions subsequent to 1537.

38. Apollonius (Perga, today Antalia, 262 a.C.— Alexandria, 190 a.C.), *The conic sections*.
39. Archimedes, *Mechanics*, and also *The quadrature of parabola*.
40. This estimate of Earth's radius is correct by 4% compared to the currently accepted size of 6371 km. The most powerful cannons used in the First World War had a range of about 14 km. The current "super-cannons" reach up to about 200 km of range.
41. On the third day, although ambiguously, Galilei seems to have said he did.
42. Particularly interesting intuition. It would be "marvelous" if the planets, "falling" from a single point, acquired the speed appropriate for their orbits. Galilei does not know the law of universal gravitation, and he does not have the physical nor the mathematical instruments necessary to carry out this calculation correctly. Here, as he often does, he says and does not say, he suggests that he knows more than he says, and indicates that someone could do the calculation—even if today we know that at the time this could not be true. Doing the calculation, we would obtain (for example by applying the virial theorem) that if the planets started from infinity they would acquire a speed equal to the orbital speed multiplied by the square root of two: not bad for an estimate purely based on the intuition and physical sense—and Galilei had a lot of either. Note that the argument is complex, and behind this complexity Galilei once again hides his heliocentric thinking.
43. The statement makes it clear that Galilei wanted to include the day of discussions on the impact (*The force of percussion*).
44. From here to the equation (20) included we move far away from the original text, for the reasons explained in note 40. Galilei presents some propositions (3-13) and derives ballistic tables that provide the maximum range and height as a function of the angle and speed of exit from the cannon's mouth; the content is implicit in the formulas we are about to demonstrate.
45. The following figure does not belong to Galilei's book.
46. For his demonstration Galilei follows the opposite path to what we are following. He first demonstrates what we demonstrate in the following paragraph, and then states that the motion of the bullet is reversible over time, and therefore the launch of a cannonball can be described by the sum of two semi-parabolas. In a letter to Mersenne Descartes greatly criticizes this demonstration, and writes that like many of Galilei's demonstrations it is "pulled out of the air".
47. Here and in the following Galilei uses the word "force" in the sense in which we use it today.
48. Aristotle, *Mechanical Problems*, 20.
49. One more reference to the additional day on the strength of percussion.
50. Aproino (1586–1638) was enrolled in the faculty of artists of the University of Padua (which included studies of astronomy, dialectics, philosophy, grammar, medicine, and rhetoric), and was a pupil of Galilei, who noticed his aptitude for physical research and associated him to his mechanical studies. After graduation, he maintained personal relations and correspondence with his teacher.
51. Antonini (1588–1616) had been a student of Galilei in Padua between 1608 and 1610, and in the following years entertained a correspondence with him. The

combination of his name with that of Aproino suggests that the experiments presented here, as well as other key experiments that Galilei described in his manuscripts, belong to the late Paduan period.

52. The balance of Galilei has been studied in detail by Ernst Mach in his fundamental treatise *Die Mechanik in ihrer Entwickelung historisch-kritisch dargestellt* [32], from which the sketch shown here is taken.

53. Although in a different context, the experimental apparatus described here is that of the instrument known today as the Atwood machine.

54. Here the posthumous publication by Galilei ends. The next four paragraphs are taken from the first and fourth of seven fragments found and published by Favaro, for some of which (in particular the second) authenticity is doubtful.

Afterword

No new edition of a work by Galilei, and in particular of the *Two New Sciences* [1, 2], can ignore the immense work done by Antonio Favaro, curator of the national edition [3] of his works. Between the late XIX century and the beginning of the XX century, among other things, Favaro established the standard version for the *Two New Sciences* by a critical examination of the many published editions, and in addition, he provided at least three major contributions.

- He analyzed in volume 72 of the Galilean manuscripts at the National Library of Florence (which essentially contains all the reports on experiments related to motion) the pages related to this book. Very often the Galilean manuscripts contain information more generous than the published text; they are what we modern experimental physicists call "logbooks".
- He reconstructed the library of Galilei [4]. This is very important to understand the origin of apocryphal citations and versions of theorems, for example attributed to Archimedes, not consistent with the Greek texts accepted today.
- He carefully reconstructed Galilei's working environment in the Paduan period, in which most of the experiments on mechanics were carried out [5, 6], including the study of the possible influence of his "human" condition: frequentations, habits, passions of a person in his youth.

Therefore, I had no doubts about the starting point of a modern edition: I chose the edition of Favaro [3], with the inclusion of reproductions of handwritten notes on a few selected topics.

The process that brought Galilei to his analysis of statics and motion has been long, transversal, and subject to interruptions. During the period of his studies in Pisa, from 1580 until 1585, Galilei observed experimentally the isochronism of pendulum oscillations. Hydrostatic and balance studies continued in the Florentine period and in the Pisan teaching period, before he moved to Padua in 1592. The Paduan lessons on mechanics (known as *Le Mecaniche*), for which we have several apocryphal manuscript translations [7] disagreeing among themselves in many respects, and one printed posthumously by a pupil who inserted many mistakes [8], indicate the

© The Editor(s) (if applicable) and The Author(s), under exclusive license to Springer
Nature Switzerland AG 2021
A. De Angelis, *Galileo Galilei's "Two New Sciences"*,
History of Physics, https://doi.org/10.1007/978-3-030-71952-4

evolution of his thought; we know that Galilei had many wrong opinions, that he will criticize in the *Two New Sciences* published in 1638. Father Marin Mersenne, a great French mathematician, translated his lessons and his works even before they were printed in Italy (we owe him the oldest version of the *Mechanics* [9]). Mersenne had also a close correspondence with Descartes, who was very critical of Galilei's methods. The post-Paduan phase is analyzed accurately by Drake [10].

For the translation of the first four days, part of the text is based (for what attains the language) on a revision of the public domain work by Crew and De Salvio [11]. The translation of the additional day is, instead, original. Copyright considerations made it impossible to revise the translation by Drake.

As for the translation into an "algebraic" language, there was no literature before this work of mine. The first to attempt an operation of this type was Mersenne, the year after the publication of Galilei's work, with a very original and particular work—Mersenne's personality shines through the work [12]. Among the many editions, the one in Spanish edited in 1945 by the Argentines José San Roman Villasante and Teófilo Isnardi in Buenos Aires [13] and the French one edited in 1970 by Maurice Clavelin [14] in Paris seem to me to be the deepest and those that introduce "added value". Another edition that I found very rich in its originality and difference from the others, particularly for an engineer's point of view on Galilei's work, is the almost unknown one by Pierini [15], who works outside the academy. I have also sometimes consulted the version of Carugo and Geymonat [16] and by Giusti [17], and the partial one by Shea and Davie [18], but I have used them less than the previous ones. On two specific topics I used the works of Benvenuto [21] and Di Pasquale [22], comparing the Galilean analysis on the resistance of materials with the modern vision, and that of Maracchia [23], comparing the Galilean demonstrations with those of Archimedes and analyzes the apocryphal literature. For the techniques of measurement of time, I used the reconstructions made by Drake [10], Vergara Caffarelli [24], Settle [25], Lepschy and Viaro [26], Bellone [27], Galluzzi [28]. Precious methodological considerations can be found in the articles by Plonitsky & Reed [29], Wallace [30] and Drake [31].

Where possible and unless otherwise specified, the figures are photographic scans of the most ancient editions, done *ex novo* in collaboration with the Biblioteca Nazionale Centrale di Firenze with the most modern techniques: from the original edition published in 1638 [1] for the first four days, and for the additional day from the 1718 edition [2], where it appears for the first time. On the basis of the comparison with the handwritten notes and in accordance with Favaro's opinion, these seem generally drawn by the Author himself (I disagree with Favaro regarding the fourth day and the additional day, for which the hand of students seems visible, consistent with the progression of Galilei's blindness). In a very limited number of cases, I have used new figures, or I have slightly modified the original ones, and these modifications are always stressed in the endnotes. I added one figure made by Ernst Mach in his fundamental work [32]. I can see that, thanks to technology and especially to the extreme care of Dr. Susanna Pelle from the Biblioteca Nazionale Centrale di Firenze and of Dr. Enrico Casadei from Codice Edizioni, in this work the reproduction of figures is more faithful than in any edition after the original one by Galilei.

I based myself on bilingual texts regarding Aristotle's works *Physics* [33] and *On the Heavens* [34], and regarding those by Archimedes [35]. For Archimedes, I also used the English translation of Heath [36]. The quotations of Aristotle's fragments refer to the numbering of Bekker [37], except for the *Mechanical Problems* for which I refer to the question number.

As I said I wanted to write a "modern" version, and sometimes I had to make choices. An example, I hope harmless, is that in my version the characters are not at all ceremonious: they behave as modern friends would. From the mathematical point of view, I was more conservative than from the lexical one, and I don't use mathematical tools unknown at the time of Galilei (in particular I don't use integral calculus and I don't explicitly use differential calculus, and not even the sum of infinite series); I use however some mathematical tools that were invented at the time as Galilei by the French school (Descartes, Mersenne), such as algebraic formulation and some analytical geometry. My notation is, of course, modern.

Omissions from the original text are few and explicitly reported. They refer to aspects that appear redundant in consideration of the translation into formulas; the reader can however express his opinion based on the final notes, and possibly consult the original version. There are very few additions: three figures, and a remarkable demonstration that follows directly from the algebraic formulation of the motion of the projectile on the fourth day. I stressed these modifications; for the latter in particular I had some doubts, because it simplifies a complex mental process of Galilei that Descartes criticized severely. I hope I have expressed my doubts to the reader, but I had to make choices. On other small additions, made by Favaro on the basis of the notes in the margin of the personal copy of Galilei, I did not make detailed comments, because it seems to me that they do not affect the substance of the work.

The day on the force of percussion, which I called Additional Day, deserves a special discussion [3, 10, 38, 39]. As said this day was published posthumously [2], and therefore it is difficult to understand if some fragments are apocryphal . Favaro separates a part almost certainly due to Galilei from some fragments that have a different degree of attributability (some, like the second, seem clearly added by students who knew post-Galilei evolutions of physics). I made a choice using my experience as an experimental physicist and I fully inserted the part attributed by Favaro to Galilei, conservatively adding only a small part of the fragments, which I feel I can say only contained statements that could have belonged to the Author on the basis of his knowledge and of his discoveries. A deep analysis of the work and in particular of the additional day, that Favaro calls the "sixth day" referring to the Florentine edition of 1718 (a "fifth day" in which Galilei defines proportionality in a more intuitive way compared to the 5th book of Euclid's *Elements of Geometry* had also appeared in the edition of 1718), is due to Mach [32]. Again this is clearly specified in the final notes.

Biographical notes are taken from Favaro's [40] and Camerota's [41] chronologies; I read with great pleasure the original and inspiring biographies written by

Heilbron [42], by Wootton [43] and by Greco [44], the book by Bucciantini [45], and the anthology by Finocchiaro [46].

I hope readers will send me comments and opinions.

Finally, as I wrote in the introduction, to make Galilei easy to read, I benefited from the collaboration of many friends, and I would like to thank them. The errors in reading their contributions are all attributable to me, who will always be very grateful to them.

Acknowledgements

Colleagues and friends Ugo Amaldi, Cesare Barbieri, Michele Bellone, Giacomo Bonnoli, Luisa Bonolis, Giovanni Busetto, Michele Camerota, Giovanna D'Agostino, Stefania De Angelis Williams, Michela and Nicola De Maria, Mosè Mariotti, Alessandro Pascolini, Nando Patat, Riccardo Rando, Maria Luisa Rischitelli, Antonio Saggion, Luigi Secco, Andrea, Nadia and Valeria Sitzia, Paolo Spinelli, Marco Tavani, Rossana Vermiglio, Jeff Wyss, contributed to the demonstrations and to the "quality control" of the text. Alessandro Bettini and Gianni Comini provided me with ideas and information that I would have hardly come up with on my own. Francesco De Stefano helped me in improving the mathematical exposition in several points. I appreciated a lot discussing with Telmo Pievani, who encouraged me and gave me some new ideas. The support from the editor, Marina Forlizzi, was, as usual, kind and effective. I would have liked to write a better book to do them much honor; I hope that they will understand that the defects in my text are not due to unwillingness but rather to my limitations, and that they will forgive me.

Part of this work has been written in Palazzo del Torso in Udine, formerly residence of that Daniele Antonini who contributed to the experiments quoted in the Additional Day and now the home of the "Centre International des Sciences Mécaniques". I finally thank the Biblioteca Nazionale Centrale and the Galileo Museum in Florence; the Center for History of the University, the University Library and the "Bruno Rossi" Library of the Department of Physics and Astronomy "Galileo Galilei" of the University of Padua; the Scientific Library of the University of Udine.

Chronology of Galilei's times

1543 Copernicus publishes *De revolutionibus orbium coelestium*, in which he affirms, on an astronomical basis, the heliocentric system.

1546 Tycho Brahe is born in Knutstorp in Scania (Denmark, today southern Sweden). He will improve the accuracy of astronomical observations, facilitating the subsequent development of physical theories.

1548 Filippo Bruno, known as Giordano, is born in Nola.

1563 The Council of Trento ends, affirming the values of the Roman Catholic Church against the diffusion of Protestantism, and initiating the Counter-Reform.

1564 Galileo Galilei is born in Pisa (Tuscany), the eldest of seven children of Giulia Ammannati and Vincenzo, luthier and theorist of harmony but also experienced in the field of music, merchant for economic hardship. The surname originated from the ancestor Galileo Bonaiuti, an illustrious physician born in 1370 and buried in Santa Croce.

1571 Johannes von Kepler is born in Weil der Stadt in Germany.

1572 A rare and spectacular astronomical event illuminates the sky: a Galactic supernova (the so-called "Tycho supernova", studied in detail by Brahe). For a couple of months an object brighter than all planets appears in the sky and then disappears from view. Galilei is eight years old.

1574 The Galilei family moves to Florence.

1581 Vincenzo Galilei publishes the treaty of harmony *Della musica antica et della moderna* (*Of ancient and modern music*).

1581 Galilei enrolls in the School of Medicine in Pisa, but he quickly switches to mathematics.

1584 Giordano Bruno publishes in London a trilogy of physical-cosmological dialogues: *De la causa, Principio Et Uno* (*On Cause, Principle and Unity*), *La cena delle ceneri* (*The Ash Wednesday Supper*), *De l'infinito, universo e mondi* (*Of the Infinite Universe and Worlds*). The treatises present, among other things, arguments in favor of heliocentrism.

© The Editor(s) (if applicable) and The Author(s), under exclusive license to Springer Nature Switzerland AG 2021
A. De Angelis, *Galileo Galilei's "Two New Sciences"*,
History of Physics, https://doi.org/10.1007/978-3-030-71952-4

1587   Without a degree, after discovering the isochronism of pendulum oscillations, Galilei competes for a chair of mathematics at the University of Bologna, but Gianantonio Magini from Padua ia preferred to him.

1589   As a reader in Mathematics in Pisa, he studies gravity and motion, and introduces new methodologies in the description of nature.

1591   Giordano Bruno publishes in Frankfurt the treatises *De minimo* (*On the Minimum*), *De monade* (*On the Monad*), and *De immenso* (*On the Vastness*), affirming the Copernican theory with new arguments.

1591   Vincenzo Galilei dies and Galileo shoulders the domestic and financial responsibility of the family.

1592   The Veneto Senate appoints Galileo professor of mathematics at the University of Padua (the university of the Republic of Venice). The salary, 180 *ducati* a year,[a] was three times larger than what he had received at Pisa. At the selection also Giordano Bruno and Magini participate as candidates.

1592   Galilei moves to Padua, studies mechanics, invents the geometric and military compass, writes notes for students. His annual teaching duties consist of sixty half-hour lessons. In his Padua house lives the craftsman Marcantonio Mazzoleni, who helps him in the construction of his instruments and experiments. Frequently Galilei also houses students.

1596   René Descartes (Cartesius) is born in La Haye en Touraine.

1597   In a letter to Kepler, Galilei professes himself as a Copernican.

1600   His daughter Virginia is born from the Venetian cohabitant Marina Gamba; Galilei is 36 years old. Daughter Livia is born the following year and son Vincenzo in 1606.

1600   Giordano Bruno is condemned as a heretic and burned at the stake in Rome.

1601   Tycho Brahe dies. At the age of 30, Kepler succeeds him as an astronomer of the emperor of the Holy Roman Empire in Prague.

1604   A new Galactic supernova appears in the sky, a little less bright than that seen by Galilei child (but still more than all planets except Venus); Galilei devotes three public lectures to the event. This has been the last of the seven Galactic supernovae in the history of humanity for which a written record is available: 185 A.D., 393, 1006, 1054 (the remnant of which is the Crab nebula), 1181, 1572 (dubbed Tycho supernova), and 1604 (Kepler supernova).

1605   Galilei spends the summer in Florence where he teaches mathematics to Cosimo dei Medici, son of the Grand Duke of Tuscany.

1606   Galilei is enrolled in the Accademia della Crusca.

1607   Invention of the thermoscope, progenitor of the thermometer.

1608   Galilei's brother Michelangelo, a luthier, became a resident musician at the court of Bavaria.

---

[a] It is difficult to compare the value of money in different eras, but to give an idea of how much this salary meant, the typical cost of staying in a hotel for a week with one's horse in the Venetian countryside was of a *ducato*, including food for both.

1609 Kepler publishes a treatise containing his first two laws; in particular he shows that the orbits of the planets are not circular. Cosimo de' Medici, upon his father's death, becomes Grand Duke of Tuscany with the name of Cosimo II.

1609 Galilei receives the drawings of a telescope from Holland; he perfects the instrument and presents it to the Doge of Venice on the bell tower of San Marco. He receives immediately a lifetime contract as a teacher with a substantial salary increase. The final salary of Galilei in Venice, salary that he never will cash, is 1000 *ducati* per year.

1610 Galilei, looking at the Milky Way and studying its structure, discovers the satellites of Jupiter, called Medicean in honor of the Grand Duke of Tuscany, and publishes the *Sidereus Nuncius* (*Starry Messenger* or *Message from the stars*: the Latin is ambiguous) in Venice; in this book he describes for the first time astronomical observations made with a scientific instrument (the telescope).

1610 At the age of 46 Galilei is appointed court mathematician and philosopher to the Medici Dukes of Tuscany in Florence, with the same salary he had in Padua, but without any obligation to teach or reside. He starts his new job on the first of September. He observes sunspots, and he discovers the phases of Venus and the rings of Saturn.

1611 Galilei explains in Rome his discoveries. He is received by Pope Paul V and is admitted to the Accademia dei Lincei.

1612 Marina Gamba dies. Galilei entrusts his daughters to his mother and his son to a governess.

1614 In the church of Santa Maria Novella, the Dominican Tommaso Caccini attacks Galilei for his "false interpretations of Scripture".

1615 Galilei goes to Rome to defend his cosmological interpretation.

1615 Galilei is denounced to the Holy Office.

1616 The Holy Office condemns the Copernican theory and warns Galilei not to support it.

1617 Galilei moves to the villa of Bellosguardo above Florence, frequented by scholars and disciples.

1617 Galilei's daughters, Virginia and Livia, take vows and become nuns respectively with the name of Maria Celeste and Arcangela.

1619 Galilei pupil and assistant Mario Guiducci published the *Discourses on Comets*; a controversy starts between Galilei and the Jesuit Orazio Grassi on the interpretation of the comets phenomena.

1619 Kepler publishes his third law.

1620 Death of Galilei's mother.

1621 Grand Duke Cosimo II, protector of Galilei, dies; the eleven-year-old Ferdinando II succeeds him, under the protection of his mother Magdalene of Austria.

1623 Maffeo Barberini, Galilei's long time friend and supporter, is elected pope with the name of Urban VIII; Galilei publishes *Il Saggiatore*.

1628 Galilei begins writing a treatise on the Universe, which will then become the book known today as *Dialogue over the two chief world systems*.

1629 Galilei's son Vincenzo marries Sestilia Bocchineri from Prato and grandchild
      Galileo is born.
1630 Kepler dies; Galilei goes to Rome to get permission to publish his *Dialogue*.
1631 Galilei moves to the villa "Il gioiello" in Arcetri: his daughter Maria Celeste,
      living in the nearby convent, takes care of him. Galilei's brother Michelangelo,
      with whom he had some friction, dies at the age of 56. Galilei will take the
      economic care of Michelangelo's eight children.
1632 At the age of 68 Galilei published in Florence, with the ecclesiastical *impri-
      matur,* the *Dialogue*. In the same year, the Church regrets the decision and
      orders the printer to suspend selling of the book and forbids Galilei to spread
      it. The first print run of over 500 copies was anyway already sold out and the
      work was circulating around Europe.
1633 Galilei is summoned to appear before the Inquisition. Trial and condemnation
      of the Holy Office; Galilei is forced to recant his ideas. Confined at Villa
      Medici in Rome and then in Siena at the home of Archbishop Piccolomini. At
      Christmas got confinement in Arcetri, near the convent that housed his daughter
      Maria Celeste, who assisted him. The writing of the *Two New Sciences* begins.
1633 Descartes, who had written the *Traité du monde et de lumiere* (in which he
      supported the heliocentric theory), renounces to publish it. The innovative
      mathematical methods of Descartes are discussed in the school of Father Marin
      Mersenne in Paris together with the lecture notes on mechanics by Galilei, that
      Mersenne translated into French in 1634 well before the Italian publication.
1634 Death of Maria Celeste.
1636 Galilei is becoming blind and struggling to read, write, and draw.
1637 Descartes publishes in Leiden his *Discours de la méthode*; his essay on
      geometry, with the introduction of analytical geometry, revolutionize the
      mathematical-geometric way of thinking.
1638 At 74 Galilei publishes the *Discourses and mathematical demonstrations about
      Two New Sciences*. Received Milton's visit.
1639 Galilei health deteriorates: he is sick and completely blind. His student Vin-
      cenzo Viviani is admitted assisting him.
1641 His student Evangelista Torricelli is also allowed to assist him.

1642 Galilei dies at the age of seventy-eight. His body is placed in the bell tower
      of Santa Croce in Florence, but not next to his father inside the Basilica, as
      Galilei would have liked.

1642 Birth of Isaac Newton.
1676 The Danish astronomer Ole Rømer performs the first measurement of the
      speed of light, using the times of occultation of one of the Medicean satellites
      of Jupiter, which depend on the fact that the satellite approaches the Earth or
      moves away from it. The result is correct within 30%.
1687 Newton publishes *Philosophiae Naturalis Principia Mathematica* (*Mathemat-
      ical Principles of Natural Philosophy*), which provide an extension of Galilei's
      physics and formalize many of his conjectures. He attributes to Galilei the merit
      of his first law (the principle of inertia) and of his second law.

1736 On the initiative of the last Medicean Grand Duke of Tuscany, Giangastone, the remains of Galilei are translated, together with those of Vincenzo Viviani and of a woman, probably Maria Celeste, in a monumental tomb inside the Basilica of Santa Croce.

1757 Most of the books promoting heliocentric astronomy are removed from the Index of the forbidden books; the books of Copernicus and Galilei are anyway maintained in that list.

1835 The books by Copernicus and Galilei are removed from the index of forbidden books. A long process of revision starts inside of the Church, which will lead to the rehabilitation of Galilei in 1992.

1851 With his famous pendulum experiment, Foucault shows that the geocentric hypothesis is incompatible with Newtonian mechanics.

Bibliography

1. Galileo Galilei, *Discorsi e dimostrazioni matematiche intorno a due nuove scienze attinenti alla mecanica & i movimenti locali...*, Elzevir, Leiden 1638
2. Galileo Galilei, *Discorsi..., Opere*, vol. 2, Tartini e Franchi, Firenze 1718
3. Galileo Galilei, *Discorsi...*, in Le Opere di Galileo Galilei, Edizione Nazionale, Antonio Favaro ed., vol. 8, Barbera, Firenze 1898, and following editions 1933, 1965, 1968; includes handwritten notes from the Author, and fragments
4. Antonio Favaro, *La libreria di Galileo Galilei descritta e illustrata*, Bollettino di Bibliografia e di storia delle Scienze matematiche e fisiche, XIX, 1886, pp. 219–293; *Appendice alla prima libreria di Galileo Galilei descritta e illustrata*, ibid., XX, 1887, pp. 337–372
5. Antonio Favaro, *Galileo Galilei e lo Studio di Padova*, Le Monnier, Firenze 1883
6. Antonio Favaro, *Galileo Galilei a Padova*, Antenore, Padova 1968
7. Galileo Galilei, *Le mecaniche*, probably 1598, vol. 2 of the Edizione Nazionale, Antonio Favaro ed., Barbera, Firenze 1898
8. *Della scienza mecanica e delle utilità che si traggono da gl'istromenti di quella*, opera cavata da manoscritti dell'eccellentissimo matematico Galileo Galilei dal cavalier Luca Danesi da Ravenna, Camerali, Ravenna 1649
9. *Les méchaniques de Galilée mathématicien & ingénieur du Duc de Florence*, Père Marin Mersenne ed., Guenon, Paris 1634
10. Galileo Galilei, *Two New Sciences*, Stillman Drake trad. & ed., University of Wisconsin Press, Madison 1974; 2nd ed., Walland Emerson, Toronto 1989
11. Galileo Galilei, *Dialogues Concerning Two New Sciences...*, translated from Italian and Latin into English by Henry Crew & Alfonso De Salvio, with an Introduction by Antonio Favaro, New York, 1914, and following editions 1933, 1939, 1946, 1950
12. Père Marin Mersenne, *Les nouvelles pensées de Galilée*, Guenon, Paris 1639
13. Galileo Galilei, *Dialogos acerca de Dos Nuevas Ciencias*, José San Roman Villasante trad. & ed., Losada, Buenos Aires 1945
14. Galileo Galilei, *Discours et demonstrations mathématiques concernant Deux Sciences Nouvelles*, Maurice Clavelin trad. & ed., Colin, Paris 1970
15. Galileo Galilei, *Discorsi...*, Claudio Pierini ed., Simeoni, Verona 2011
16. Galileo Galilei, *Discorsi...*, Adriano Carugo & Ludovico Geymonat ed., Boringhieri, Torino 1958; includes notes and fragments
17. Galileo Galilei, *Discorsi...*, Enrico Giusti ed., Einaudi, Torino 1990

© The Editor(s) (if applicable) and The Author(s), under exclusive license to Springer Nature Switzerland AG 2021
A. De Angelis, *Galileo Galilei's "Two New Sciences"*, History of Physics, https://doi.org/10.1007/978-3-030-71952-4

18. William R. Shea and Mark R. Davie, *Galileo Galilei Selected Writings*, Oxford University Press, Oxford 2012
19. Stephen Hawking, *On the Shoulders of Giants: The Great Works of Physics and Astronomy*, Running Press, Philadelphia 2002
20. Alexandre Koyré, *Études galiléennes*, Hermann, Paris 1939
21. Edoardo Benvenuto, *La scienza delle costruzioni nel suo sviluppo storico*, Sansoni, Firenze 1981
22. Salvatore Di Pasquale, *L'arte del costruire*, Marsilio, Venezia 1996
23. Silvio Maracchia, *Galileo e Archimede*, unpublished
24. Roberto Vergara Caffarelli, *Il laboratorio di Galileo*, self-published, Pavia 2005
25. Thomas B. Settle, *An experiment in the history of science*, Science 133 (1981) 19
26. Antonio Lepschy and Umberto Viaro, *Galileo e la misura dello spazio e del tempo*, Atti delle Celebrazioni Galileiane in Padova, LINT Trieste, 1995, p. 109
27. Enrico Bellone, unpublished lectures
28. Paolo Galluzzi, sezione sulla misura del tempo at the Museo Galileo, Firenze
29. Arkady Plonitsky and David Reed, *Discourse, Mathematics, Demonstration, and Science in Galileo's Discourses Concerning Two New Sciences*, Configurations 9 (2001) 37
30. William Wallace (1974). *Galileo and Reasoning Ex Suppositione: The Methodology of the Two New Sciences* (p. 79). Proc. of the Biennial Meeting of the Philosophy of Science Association: Boston Studies in the Philosophy of Science, Springer.
31. Stillman Drake, *Galileo's Discovery of the Law of Free Fall*, Scientific American 228#5 (1973) 84
32. Ernst Mach, *Die Mechanik in ihrer Entwickelung historisch-kritisch dargestellt* (Mechanics in its historical-critical development), Brockhaus, Leipzig 1883; translated by T. McCormack as *The Science of Mechanics*, Open Court, Chicago 1919
33. Aristotle, *Fisica*, Luigi Ruggiu transl. & ed., Rusconi, Milano 1995
34. Aristotle, *Du ciel*, Paul Moraux ed., Les belles lettres, Paris 2003
35. Archimedes, *Oeuvres*, Charles Mugler ed., Les belles lettres, Paris 1970
36. Thomas Heath, *The Works of Archimedes*, Cambridge University Press, Cambridge 1897
37. *Aristotelis Opera*, 5 voll., Immanuel Bekker ed., Academia Regia Borussica, Berlin 1831–1870
38. *Galileo in context*, Jürgen Renn ed., Cambridge University Press, Cambridge 2001
39. Roberto Vergara Caffarelli, *Il principio d'inerzia negli ultimi scritti di Galileo*, in *A reconstruction of 50 years of experiments and discoveries*, SIF-Springer, Heidelberg 2009
40. Antonio Favaro, *Cronologia Galileiana*, R. Accademia di Scienze Lettere ed Arti in Padova, 1891
41. Michele Camerota, *Galileo Galilei*, Corriere della Sera, Milano 2019
42. John L. Heilbron, *Galileo*, Oxford University Press, Oxford 2010
43. David Wootton, *Galileo: watcher of the skies*, Yale University Press, New Haven 2010
44. Pietro Greco, *Galileo Galilei, the Tuscan Artist*, Springer Nature, Heidelberg 2018
45. Massimo Bucciantini, *Galileo e Keplero*, Einaudi, Torino 2007
46. Maurice Finocchiaro, *The Essential Galileo*, Hackett, Indianapolis 2018
47. Luca Valerio, *De centro gravitatis solidorum*, Bonfadino, Roma 1604
48. Galileo Galilei, *Sidereus Nuncius*, Baglioni, Venezia 1610
49. Galileo Galilei, *Discorso intorno alle cose che stanno in su l'acqua, o che in quella si muovono*, Giunti, Firenze 1612
50. Galileo Galilei, *Il Saggiatore*, Mascardi, Roma 1623
51. Galileo Galilei, *Dialogo sopra i due massimi sistemi del mondo*, Landini, Firenze 1632
52. Isaac Newton, *Philosophiae Naturalis Principia Mathematica*, Streater, London 1687

Printed in the United States
by Baker & Taylor Publisher Services